Nanocomposites of Polymers and Inorganic Particles

Nanocomposites of Polymers and Inorganic Particles

Editors

Marinella Striccoli
Annamaria Panniello
Roberto Comparelli

MDPI • Basel • Beijing • Wuhan • Barcelona • Belgrade • Manchester • Tokyo • Cluj • Tianjin

MDPI

Editors

Marinella Striccoli	Annamaria Panniello	Roberto Comparelli
IPCF - Institute for Physical and Chemical Processes	IPCF - Institute for Physical and Chemical Processes	IPCF - Institute for Physical and Chemical Processes
CNR	CNR	CNR
Bari	Bari	Bari
Italy	Italy	Italy

Editorial Office
MDPI
St. Alban-Anlage 66
4052 Basel, Switzerland

This is a reprint of articles from the Special Issue published online in the open access journal *Molecules* (ISSN 1420-3049) (available at: www.mdpi.com/journal/molecules/special_issues/Nanocomposites-Particles).

For citation purposes, cite each article independently as indicated on the article page online and as indicated below:

LastName, A.A.; LastName, B.B.; LastName, C.C. Article Title. *Journal Name* **Year**, *Volume Number*, Page Range.

ISBN 978-3-0365-1760-5 (Hbk)
ISBN 978-3-0365-1759-9 (PDF)

Contents

About the Editors vii

Preface to "Nanocomposites of Polymers and Inorganic Particles" ix

Maria Eriksson, Joris Hamers, Ton Peijs and Han Goossens
The Influence of Graft Length and Density on Dispersion, Crystallisation and Rheology of Poly(-caprolactone)/Silica Nanocomposites
Reprinted from: *Molecules* **2019**, *24*, 2106, doi:10.3390/molecules24112106 1

Polina Demina, Natalya Arkharova, Ilya Asharchuk, Kirill Khaydukov, Denis Karimov, Vasilina Rocheva, Andrey Nechaev, Yuriy Grigoriev, Alla Generalova and Evgeny Khaydukov
Polymerization Assisted by Upconversion Nanoparticles under NIR Light
Reprinted from: *Molecules* **2019**, *24*, 2476, doi:10.3390/molecules24132476 21

Alojz Anžlovar, Mateja Primožič, Iztok Švab, Maja Leitgeb, Željko Knez and Ema Žagar
Polyolefin/ZnO Composites Prepared by Melt Processing
Reprinted from: *Molecules* **2019**, *24*, 2432, doi:10.3390/molecules24132432 33

Suzana Natour, Anat Levi-Zada and Raed Abu-Reziq
Magnetic Polyurea Nano-Capsules Synthesized via Interfacial Polymerization in Inverse Nano-Emulsion
Reprinted from: *Molecules* **2019**, *24*, 2663, doi:10.3390/molecules24142663 47

Lisha Ai, Yejing Wang, Gang Tao, Ping Zhao, Ahmad Umar, Peng Wang and Huawei He
Polydopamine-Based Surface Modification of ZnO Nanoparticles on Sericin/Polyvinyl Alcohol Composite Film for Antibacterial Application
Reprinted from: *Molecules* **2019**, *24*, 503, doi:10.3390/molecules24030503 63

Yejing Wang, Rui Cai, Gang Tao, Peng Wang, Hua Zuo, Ping Zhao, Ahmad Umar and Huawei He
A Novel AgNPs/Sericin/Agar Film with Enhanced Mechanical Property and Antibacterial Capability
Reprinted from: *Molecules* **2018**, *23*, 1821, doi:10.3390/molecules23071821 75

Gerald Singer, Philipp Siedlaczek, Gerhard Sinn, Patrick H. Kirner, Reinhard Schuller, Roman Wan-Wendner and Helga C. Lichtenegger
Vacuum Casting and Mechanical Characterization of Nanocomposites from Epoxy and Oxidized Multi-Walled Carbon Nanotubes
Reprinted from: *Molecules* **2019**, *24*, 510, doi:10.3390/molecules24030510 89

Paula A. Zapata, Humberto Palza, Boris Díaz, Andrea Armijo, Francesca Sepúlveda, J. Andrés Ortiz, Maria Paz Ramírez and Claudio Oyarzún
Effect of $CaCO_3$ Nanoparticles on the Mechanical and Photo-Degradation Properties of LDPE
Reprinted from: *Molecules* **2018**, *24*, 126, doi:10.3390/molecules24010126 99

Jong-Min Lim and Sehee Jeong
Fabrication of Spherical Titania Inverse Opal Structures Using Electro-Hydrodynamic Atomization
Reprinted from: *Molecules* **2019**, *24*, 3905, doi:10.3390/molecules24213905 111

About the Editors

Marinella Striccoli

Dr. Marinella Striccoli is a senior researcher with the National Council of Researches—Institute for Physical and Chemical Processes (CNR IPCF) in Bari, Italy. Her expertise covers the optical and morphological characterization of colloidal nanomaterials and hybrid organic–inorganic nanostructures as well as nanocomposite materials to be used in optoelectronic and energy conversion applications. She has actively worked as a PI in several European Projects (H2020 FET, large 7FP EU Projects) and in several national and regional projects. In addition, she collaborated in research activities in the field of material science and nanostructures. She is the co-author of more than 160 peer-reviewed papers, 1 patent, and 7 book chapters.

Annamaria Panniello

Dr. Annamaria Panniello obtained her Ph.D. in the Chemistry of Innovative Materials at the University of Bari in 2010. Since December 2017, she has been a scientist at CNR-IPCF. Her scientific research is focused on material chemistry, aiming at the design and fabrication of oxide, semiconductor, and carbon-based nanoparticles (NPs) and on their post-synthesis processing by surface functionalization; immobilization onto proper substrates; and/or their incorporation in matrices, such as polymers, epoxy resists, and ionic liquids. Her expertise includes NP synthesis and their optical, morphological characterization; NPs' 2D self-organization; the preparation of nanocomposite materials for photonic, optoelectronic, and photo-catalytic applications; and energy storage. She is and has been actively involved in the research activities of EU projects (6th and 7th FP, and Horizon2020), bilateral programs, and Italian Projects (PRIN and PON). She has co-authored 30 papers, 2 book chapters, and several contributions to conferences.

Roberto Comparelli

Dr. Roberto Comparelli (Ph.D. in the Chemistry of Innovative Materials) is a senior researcher at CNR-IPCF. His expertise covers nanocrystal (NC) synthesis by wet chemistry (photoactive or magnetic oxides, II-VI semiconductors, and metals), characterization (TEM, SEM, AFM, FT-IR, UV-Vis-NIR, and PL), and their surface engineering. He is interested in NC incorporation in polymer matrices, applications in optoelectronic, self-assembly, and biological and environmental fields. In particular, he has a strong background in the application of photoactive NCs in the degradation of organic/inorganic pollutants in water and gas matrices and in the preparation and characterization of NC-based self-cleaning coatings. He has been involved in several EU and Italian projects. He has co-authored over 130 papers and more than 150 contributions to conferences with invited talks.

Preface to "Nanocomposites of Polymers and Inorganic Particles"

In the last few years, significant efforts have been devoted to designing, fabricating, and exploit nanocomposite materials based on inorganic nanoparticles incorporated in a polymer matrix [1–5]. The extraordinary interest in such materials relies on the large range of properties that can arise from the synergic combination of the features of nanoparticles (NPs) and the host polymer. Indeed, the original size-dependent physical and chemical properties of nanomaterials (semiconducting, metals, oxides, and magnetic NPs) combined with the high processability, the defined chemistry, and the morphology of polymers and block copolymers finally turn out to be innovative materials with high technological impact in a variety of advanced application in photonic, optoelectronics, sensing, environmental, energy conversion, biological, and biomedical fields [6–11].

The contributions to this Special Issue cover all of the specific aspects of this topic, ranging from preparatory approaches, functionalization strategies of NPs and polymers, processing and integration of nanocomposites in additive manufacturing materials, and technological methodologies to obtain functional multiphase materials for advanced applications.

In composites, the use of fillers at the nanometric scale allows for taking advantage of the very small size, the extended surface area, and the tunability of the interpaticle distance as a function of the filler loading compared to conventional materials based on micro-sized particles. The use of such nanofillers can improve the stiffness of polymeric hosts and prevent the material from cracks delaying the natural failure of the material. However, the homogeneous dispersion of NPs within the polymeric host is not a straightforward outcome and is crucial for good preservation of the mechanical, morphological, and physicochemical properties of the nanocomposite, as the occurrence of micro-aggregation can catalyze the mechanical stress, reduce the effective surface area, and finally worsen the mechanical and rheological properties of the multiphase material. Thus, the interactions among NPs, and between NPs and the polymer matrix needs to be investigated and optimized to fabricate effective nanocomposite materials for advanced applications. In order to tune these interactions and then to disperse the fillers in a polymer matrix, grafting of the polymer to the nanoparticle surface is a technique often employed, able to improve the affinity between fillers and polymer matrices. In the first contribution to this Special Issue [12], the influence of graft length and graft density on the state of dispersion, crystallization, and rheological properties of poly(-caprolactone) (PCL)/silica (SiO_2) nanocomposites were studied. Specifically, controlling the length and density of the grafted silica, different states of NP dispersion were found, ranging from small spherical aggregates to sheet-like microstructures depending on the matrix-to-graft molecular weight ratios.

Among the several methods for nanocomposite preparation, Demina and coworkers [13] developed an original approach based on near infrared light-activated photopolymerization of nanocomposites containing luminescent lanthanide-doped up-conversion NPs, effective both in oligomer bulk and on the NP surface in aqueous dispersion. In [14], interfacial polymerization in inverse (water-in-oil) nanoemulsion was exploited to prepare polyurea nanocapsules containing ionic liquid-modified magnetite NPs as multiphase systems with potential application in catalysis and biomedicine, such as for the targeted delivery of hydrophilic drugs. Possible walkthroughs for obtaining polymer nanocomposites with homogeneously dispersed inorganic nanoparticles, limiting the phase segregation and micro-aggregation phenomena, are chemical modification of the nanoparticle surface, proper functionalization of the polymer side-chains, or in situ synthesis of the NPs inside the host matrix. In order to introduce well-defined functionalities into polyolefin matrices, specifically UV absorption and antibacterial activity, Anžlovar et al. obtained nanocomposites through the deposition of surface-modified ZnO NPs on polyolefin granules and subsequent melt processing [15].

Besides great efforts devoted to the development of novel and original synthetic strategies to gain nanocomposite systems, a significant issue in such a intriguing field is the investigation of the mechanical, electrical, thermal, magnetic, and optical properties arising in the final composite, where the material performance has successfully benefited from the combination of peculiar features of all of the components constituting the system. An in-depth study of the mechanical properties and performance test of an epoxy resin matrix reinforced by the addition of "green"-synthesized oxidized MWCNTs is reported in [16]. The authors applied a vacuum casting approach for different geometries to limit the increase in viscosity resulting from adding CNTs to the epoxy resin; performing various mechanical tests that show improved fracture toughness, bending, and tensile properties with the addition of MWCTs; and finally, analyzing the strengthening mechanisms by SEM and in situ imaging by digital image correlation (DIC) of the fracture surface.

A further relevant issue that needs to be resolved concerns the degradation properties of plastic materials and the environmental stability of polymer nanocomposites. Indeed, achieving the biodegradation of commercial plastics is an enormous environmental challenge due to the increased social demand for higher sustainability processes. The addition of nanofillers often improves the mechanical behavior of the polymeric host, although it sometimes can accelerate photodegradation processes of the matrix. In [17], $CaCO_3$ NPs were synthesized and modified by oleic acid to improve their interaction with a low-density polyethylene (LDPE) matrix. The effect of photoaging under UV irradiation on the structural (crystallinity percentage, c), chemical (carbonyl index, CI), and mechanical (Young's modulus) properties of composites were studied and compared with those of the pure polymer to assess if the presence of $CaCO_3$ NPs accelerates the photodegradation of the LDPE.

Examples of nanofillers that strengthen the mechanical behavior of polymeric matrices are very common in the panorama of nanocomposite materials. Besides such beneficial effects, the introduction of inorganic nanoparticles within polymeric chains can also induce specific and well-defined abilities, thus allowing for their use in widespread potential applications. Wang et al. [18] proposed a novel nanocomposite film with enhanced mechanical performance and antimicrobial properties for potential biomedical applications. Specifically, a multifunctional nanocomposite obtained by the in situ growth of Ag NPs on the surface of sericin/agar film showed high mechanical property, hydrophilicity, hygroscopicity, and stability and finally demonstrated an excellent antibacterial activity against *E. coli* and *S. aureus*. An additional example of the preparation of silk sericin-based nanocomposites and its application as antibacterial material was proposed by Ai et al. [19]. In their work, the authors reinforced the mechanical properties of sericin by blending with PVA and by grafting ZnO NPs with the support of polydopamine (PDA). The ZnO NP-added PDA-SS/PVA films were characterized by improved mechanical stability, tensile strength, and elongation at break and possessed excellent hydrophilicity and swellability, demonstrating a remarkable antimicrobial activity against *E. coli* and *S. aureus* to thus be effectively exploited as active nanocomposite film in antibacterial biomaterial applications.

The last contribution to this Special Issue reported on the preparation of spherical poly (styrene-co-(2-hydroxyethyl methacrylate)) (PS/HEMA) opal structures and spherical TiO_2 inverse opal structures by electro-hydrodynamic atomization [20]. Specifically, titania NPs of relatively small size were assembled at the interstitial site of PS/HEMA NP, resulting in a spherical opal composite structure with potential application in a widespread area, such as reflective mode display, photo catalysis, solar cell electrode materials, and analytical systems.

These contributions are not exhaustive but represent an updated panorama of some of the infinite possibilities in the field of nanocomposite materials and can be an inspiration for new and advanced challenges in their preparation and application in different technological fields.

Funding: This research received no external funding.

Acknowledgments: The guest editors thank all of the authors that have contributed to this Special Issue and all of the reviewers for their evaluation of the submitted articles.

Conflicts of Interest: The authors declare no conflicts of interest.

References

[1] Mourdikoudis, S., Kostopoulou, A., LaGrow, A. P., Magnetic Nanoparticle Composites: Synergistic Effects and Applications. Adv. Sci. 2021, 8, 2004951, doi: https://doi.org/10.1002/advs.202004951

[2] Nandihalli, N.; Liu, C.-J.; Mori, T. Polymer based thermoelectric nanocomposite materials and devices: Fabrication and characteristics. Nano Energy 2020, 78, 105186, doi: https://doi.org/10.1016/j.nanoen.2020.105186.

[3] Loste, J.; Lopez-Cuesta, J.-M.; Billon, L.; Garay, H.; Save, M. Transparent polymer nanocomposites: An overview on their synthesis and advanced properties. Prog. Polym. Sci. 2019, 89, 133-158, doi: https://doi.org/10.1016/j.progpolymsci.2018.10.003.

[4] Khalid, K.; Tan, X.; Mohd Zaid, H.F.; Tao, Y.; Lye Chew, C.; Chu, D.-T.; Lam, M.K.; Ho, Y.-C.; Lim, J.W.; Chin Wei, L., Advanced in developmental organic and inorganic nanomaterial: a review. Bioengineered 2020, 11, 328-355, doi: 10.1080/21655979.2020.1736240.

[5] Pourhashem, S.; Saba, F.; Duan, J.; Rashidi, A.; Guan, F.; Nezhad, E.G.; Hou, B., Polymer/Inorganic nanocomposite coatings with superior corrosion protection performance: A review. Journal of Industrial and Engineering Chemistry 2020, 88, 29-57, doi: https://doi.org/10.1016/j.jiec.2020.04.029.

[6] Melinte, V.; Stroea, L.; Chibac-Scutaru, A.L., Polymer Nanocomposites for Photocatalytic Applications. Catalysts 2019, 9, doi: 10.3390/catal9120986.

[7] Liu, S.-W.; Wang, L.; Lin, M.; Liu, Y.; Zhang, L.-N.; Zhang, H., Tumor Photothermal Therapy Employing Photothermal Inorganic Nanoparticles/Polymers Nanocomposites. Chinese Journal of Polymer Science 2019, 37, 115-128, doi: 10.1007/s10118-019-2193-4.

[8] Adnan, M.M.; Tveten, E.G.; Glaum, J.; Ese, M.-H.G.; Hvidsten, S.; Glomm, W.; Einarsrud, M.-A., Epoxy-Based Nanocomposites for High-Voltage Insulation: A Review. Advanced Electronic Materials 2019, 5, 1800505, doi: https://doi.org/10.1002/aelm.201800505

[9] Surmenev, R.A.; Orlova, T.; Chernozem, R.V.; Ivanova, A.A.; Bartasyte, A.; Mathur, S.; Surmeneva, M.A., Hybrid lead-free polymer-based nanocomposites with improved piezoelectric response for biomedical energy-harvesting applications: A review. Nano Energy 2019, 62, 475-506, doi:https://doi.org/10.1016/j.nanoen.2019.04.090.

[10] Kausar, A., A review of high performance polymer nanocomposites for packaging applications in electronics and food industries. Journal of Plastic Film & Sheeting 2019, 36, 94-112,

doi: 10.1177/8756087919849459.

[11] Cantarella, M.; Impellizzeri, G.; Di Mauro, A.; Privitera, V.; Carroccio, S.C., Innovative Polymeric Hybrid Nanocomposites for Application in Photocatalysis. Polymers 2021, 13, doi: 10.3390/polym13081184.

[12] Eriksson, M.; Hamers, J.; Pe, T.; Goossens, H. The Influence of Graft Length and Density on Dispersion, Crystallisation and Rheology of Poly(-caprolactone)/Silica Nanocomposites. Molecules 2019, 24, 2106, doi:10.3390/molecules24112106.

[13] Demina, P.; Arkharova, N.; Asharchuk, I.; Khaydukov, K.; Karimov, D.; Rocheva, V.; Nechaev, A.; Grigoriev, Y.; Generalova, A.; Khaydukov, E. Polymerization Assisted by Upconversion Nanoparticles under NIR Light. Molecules 2019, 24, 2476, doi:10.3390/molecules24132476.

[14] Natour, S.; Levi-Zada, A.; Abu-Reziq, R. Magnetic Polyurea Nano-Capsules Synthesized via Interfacial Polymerization in Inverse Nano-Emulsion. Molecules 2019, 24, 2663, doi:10.3390/molecules24142663.

[15] Anžlovar, A.; Primožič, M.; Švab, I.; Leitgeb, M.; Knez, Ž.; Žagar, E. Polyolefin/ZnO Composites Prepared by Melt Processing. Molecules 2019, 24, 2432, doi:10.3390/molecules24132432.

[16] Singer, G.; Siedlaczek, P.; Sinn, G.; Kirner, P.H.; Schuller, R.; Wan-Wendner, R.; Lichtenegger, H.C. Vacuum Casting and Mechanical Characterization of Nanocomposites from Epoxy and Oxidized Multi-Walled Carbon Nanotubes. Molecules 2019, 24, 510, doi:10.3390/molecules24030510.

[17] Zapata, P.A.; Palza, H.; Díaz, B.; Arm, A.; Sepúlveda, F.; Ortiz, J.A.; Ramírez, M.P.; Oyarzún, C. Effect of CaCO3 Nanoparticles on the Mechanical and Photo-Degradation Properties of LDPE. Molecules 2019, 24, 126, doi:10.3390/molecules24010126.

[18] Wang, Y.; Cai, R.; Tao, G.; Wang, P.; Zuo, H.; Zhao, P.; Umar, A.; He, H. A Novel AgNPs/Sericin/Agar Film with Enhanced Mechanical Property and Antibacterial Capability. Molecules 2018, 23, 1821, doi:10.3390/molecules23071821.

[19] Ai, L.; Wang, Y.; Tao, G.; Zhao, P.; Umar, A.; Wang, P.; He, H. Polydopamine-Based Surface Modification of ZnO Nanoparticles on Sericin/Polyvinyl Alcohol Composite Film for Antibacterial Application. Molecules 2019, 24, 503, doi:10.3390/molecules24030503.

[20] Lim, J.-M.; Jeong, S. Fabrication of Spherical Titania Inverse Opal Structures Using Electro-Hydrodynamic Atomization. Molecules 2019, 24, 3905, doi:10.3390/molecules24213905.

Marinella Striccoli, Annamaria Panniello, Roberto Comparelli

Editors

molecules

MDPI

Article

The Influence of Graft Length and Density on Dispersion, Crystallisation and Rheology of Poly(ε-caprolactone)/Silica Nanocomposites

Maria Eriksson [1], Joris Hamers [1], Ton Peijs [2,*] and Han Goossens [1,3]

[1] Laboratory of Polymer Materials, Department of Chemical Engineering and Chemistry, Eindhoven University of Technology, 5600 MB Eindhoven, The Netherlands; mailmariaeriksson@gmail.com (M.E.); jorishamers@hotmail.nl (J.H.)
[2] Materials Engineering Centre (MEC), WMG, University of Warwick, Coventry CV4 7AL, UK
[3] SABIC, Plasticslaan 1, 4612 PX Bergen op Zoom, The Netherlands; Han.Goossens@sabic.com
* Correspondence: t.peijs@warwick.ac.uk; Tel.: +44-24-7657-2659

Received: 1 May 2019; Accepted: 31 May 2019; Published: 3 June 2019

check for updates

Abstract: Different techniques of grafting polymer chains to filler surfaces are often employed to compatibilise filler and polymer matrices. In this paper the influence of graft length and graft density on the state of dispersion, crystallisation and rheological properties of poly(ε-caprolactone) (PCL)/silica (SiO_2) nanocomposites are reported. Grafted silica nanoparticles were prepared through polymerisation of PCL from the nanoparticle surface. Graft length was controlled by the reaction time, while the grafting density was controlled by the monomer-to-initiator ratio. Grafted nanoparticles were mixed with PCL of different molecular weights and the state of dispersion was assessed. Different matrix-to-graft molecular weight ratios resulted in different states of dispersion. Composites based on the higher molecular weight matrix exhibited small spherical agglomerates while the lower molecular weight matrix revealed more sheet-like microstructures. The state of dispersion was found to be relatively independent of graft length and density. Under quiescent conditions the grafts showed increased nucleation ability in the higher molecular weight PCL, while in the lower molecular weight matrix the effect was less pronounced. Rheological experiments showed an increase in viscosity with increased filler content, which was beneficial for the formation of oriented structures in shear-induced crystallisation.

Keywords: nanocomposites; polymer; silica; grafting; dispersion; rheology; crystallisation

1. Introduction

Due to their large surface-to-volume ratio and poor compatibility with most polymers, inorganic nanofillers are rather difficult to disperse in many polymer matrices. In these matrices, filler-filler interactions are favoured over polymer-filler interactions and filler agglomerates are readily formed. The advantage of nanofillers over more conventional micron-sized fillers is related to their small size, large surface area and small interparticle distances already at relative low filler loadings [1]. Well-dispersed nanofillers have, in several cases, shown to enhance the stiffness of polymer matrices without compromising ductility [2,3]. In addition, nanofillers can act as barriers for crack growth and thereby delay the formation of cracks large enough to initiate failure. For example, it has been shown that the toughness of polyamide 6 (PA6)/silica nanocomposites depends on the ligament thickness between nanoparticles and that above a critical ligament thickness no toughening effect was observed [4]. However, this is only true for well-dispersed nanofillers which are smaller than the critical crack size of the polymer matrix. Large fillers or agglomerates can act as stress concentrators and can, thus, deteriorate the mechanical properties [1,3]. Heavy agglomeration is, therefore, certainly not desired, since it reduces the effective surface area of the particles and increases the interparticle distance, thereby counteracting the nanosize effect. There are, however, properties—for example, electrical or thermal conductivity—where a certain degree of agglomeration, i.e., intimate contact between fillers is beneficial for optimal performance [5–7]. A need for controlled dispersion is, therefore, generally required in nanocomposites. In order to tune the dispersion in a given composite system, the filler-filler interactions and/or the polymer-filler interactions need to be modified.

In order to modify these interactions and, hence, the dispersion behaviour of various fillers in a polymer matrix, grafting of the same [6,8–13] or a different polymer [14,15] to the filler surface is often performed. These grafts reduce filler-filler interactions and enhance the compatibility of the filler and matrix. Hasegawa et al. [16] already showed, both experimentally and theoretically, that there is an optimum graft density for dispersing polymer grafted particles in a polymer matrix. In another interesting paper Akcora et al. [9] showed that controlled agglomeration in a nanocomposite with polystyrene (PS)-grafted silica nanoparticles in a PS matrix could be used to create different microstructures, i.e., sheets of particles, spherical aggregates, strings of particles or well-dispersed, isolated nanoparticles, depending on the ratio between the molecular weight of the matrix and the grafts and the graft density.

The influence of grafting on the properties of the obtained nanocomposites is often studied for amorphous polymers. Various groups studied the influence of grafting on the glass transition temperature [13], mechanical properties [10] and rheological properties of nanocomposites [11,12,17]. Oh and Green showed that the glass transition temperature (T_g) could be increased (long grafts) or decreased (short grafts) depending on the graft length of PS grafts in the nanocomposite, in which PS-grafted gold (Au) nanoparticles were dispersed in a PS matrix [13]. A key question to answer here is whether the length of the grafts exceed the critical entanglement molecular weight of the bulk matrix. Akcora et al. showed that for poly(methyl methacrylate) (PMMA) nanocomposites the rheological behaviour depends on the difference in molecular weight between grafts and bulk matrix [11]. Nanocomposites based on a matrix with a higher molecular weight than the grafts, showed agglomerated particles and solid-like rheological behaviour at high filler loadings. However, in matrices with a lower molecular weight than the grafts, fillers were well dispersed and no sign of a solid-like behaviour was observed, not even at the highest filler loadings.

For semi-crystalline polymers, there are fewer examples of the influence of grafting on nanocomposite properties, and even less on the importance of graft length and density. In the current study poly(ε-caprolactone) (PCL) was used as the matrix material. The chemistry of grafting PCL to silica particles [18–20], is well studied and it was shown that the addition of grafted silica particles can enhance nucleation [21] and improve mechanical properties [22] of PCL/silica nanocomposites. This nucleation effect was also observed in other PCL/nanofiller systems. For instance, Zhou et al. [23] showed that grafted carbon nanotubes have a stronger nucleating effect than their non-grafted counterparts. The effect was larger in samples with a higher grafting density and shorter grafts as compared to samples with a low grafting density and longer grafts. L'Abee et al. [24] showed that, during the in situ preparation of rubber particle-filled PCL, the peroxide used to cross-link the rubber network also caused grafting of PCL chains onto the rubber particles. Transmission electron microscopy (TEM) images of the grafted particles revealed lamellae growing from the surface of the particles, while also an increase in nucleation efficiency was observed which was shown to depend on grafting density [24]. The grafted chains also influenced the mechanical behaviour of these PCL/rubber blends.

There are generally two strategies in grafting chemistry: the "grafting-to" [22] and the "grafting-from" approach [18–20]. In the "grafting-to" approach, chains of the polymer are attached to the particle surface via a coupling reaction, while in the "grafting-from" approach the polymerisation is performed in the presence of particles and an increase of the graft length with reaction time is observed [2]. Both procedures have their obvious advantages. In the "grafting-to" approach the molecular weight and the polydispersity of the chains can be determined prior to the coupling reaction, but there are limitations to the number of chains that can be added to the particles, especially if the size of the particles is small, due to the bulky size of the polymer chains. The "grafting-from" approach offers more possibilities of controlling the grafting density by tuning the number of initiating sites on the surface of the particles. Here, the limitation is that it has been difficult to determine the exact graft length and density on the particles, which complicates the analysis of the effect of the grafting on the properties of the obtained composite materials, especially if several graft lengths are to be used.

In the current study we focus on the "grafting-from" approach of PCL from a silica (SiO_2) nanoparticle surface. This reaction is well described in the literature, both directly from the silica surface [18,19] and by using silane spacers [19,20]. First, nanoparticles with different graft lengths and densities where prepared and subsequently these particles were mixed with the PCL matrix. Properties of the resulting nanocomposites will be discussed in terms of the influence of graft length and graft density on the state of dispersion, crystallisation and rheological behaviour.

2. Experimental

2.1. Materials

Colloidal silica suspension (ORGANOSILICASOL™) in isopropanol (IPA-ST) and toluene (TOL-ST) containing 40 wt% SiO_2 nanoparticles (10–15 nm) was obtained from Nissan Chemical Industries (City, Japan). The silica particles were prepared in ethanol using the well-known Stöber method, surface modified and transferred to the mentioned solvent. ε-Caprolactone from Sigma-Sigma-Aldrich Chemie B.V. (Zwijndrecht, Netherlands) was dried on molecular sieves prior to use. Toluene, hexane and ethanol were supplied from Biosolve B.V. (Valkenswaard, Netherlands). The toluene was distilled and stored on molecular sieves before use. Furthermore, ninhydrin (ACS reagent) and stannous octoate from Sigma-Aldrich Chemie B.V., and 3-aminopropyltriethoxysilane (97%) from abcr GmbH (Karlsruhe, Germany), were used. Two grades of poly(ε-caprolactone) were used as matrices. CAPA® 6400 (M_n = 40 kg/mol) and CAPA® 6800 (M_n = 80 kg/mol) were kindly provided by Perstorp Polycaprolactones (Perstorp, Sweden). These polymers are further referred to as PCL40 and PCL80, respectively. Both polymers were used as received.

2.2. Sample Preparation

Grafting experiments: First, the silica nanoparticles were silanised with a 3-aminopropyltriethoxysilane (APTES). For the reaction 4 wt% silica in toluene solution was used in a 2:1 ratio of silane-to-hydroxyl groups. The reaction was performed at 70 °C for 20–24 h and the reaction mixture was thereafter centrifuged at 10,000 rpm for 25 min, re-dispersed and centrifuged two more times before drying for 5 h at 80 °C in vacuum in order to remove residual solvent. A ninhydrin test was performed in order to determine the presence of amine groups on the silica particles after the centrifugation and drying step. A 0.4 wt% solution of ninhydrin in water was used. One millilitre of this solution was added to approximately 2 mg of sample and the solution was heated to the boiling point for about 20 s. The presence of amines could then be verified by the appearance of a blue colour. From the difference in weight loss during thermogravimetric analysis (TGA) experiments between the reacted and unreacted particles the approximate number of amino groups per square nanometre of particle surface was calculated.

These APTES-grafted particles were subsequently used in PCL grafting reactions with a $Sn(Oct)_2$ catalyst. The reactions were performed with initiator-to-catalyst molar ratios between 1:0.33 and 1:1 in a 1 wt% solution of particles in dried toluene. The temperature was set to 100 °C and the reaction time was varied in order to vary the grafting length. Varying the initiator-to-catalyst molar ratio the grafting density could be changed. A scheme of the performed reaction is presented in Figure 1.

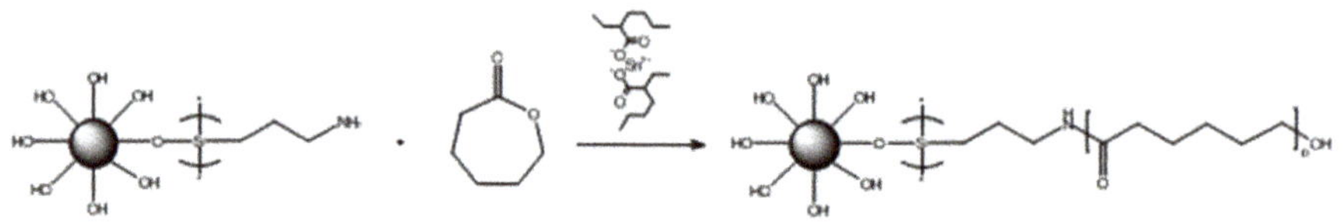

Figure 1. A scheme of the PCL grafting reaction.

Details of the different grafted nanoparticles are presented in Table 1. After the grafting reaction, the particles were separated from the reaction mixture by three successive centrifugation (3000 rpm, 30 min) and re-dispersion cycles in hexane. The complete removal of unreacted monomer was confirmed by ^{1}H NMR spectroscopy.

Table 1. Description of the different grafting experiments.

Sample Code	Number of NH (groups/nm^2)	Particles (g)	Catalyst (mg)	Caprolactone (g)	Toluene (g)	Time (h)
Short grafts	0.5–0.7	1.1	100	6.0	80	16
Medium grafts (half density)	0.6–0.8	1.1	67	5.4	90	70
Medium grafts	0.5–0.8	1.1	122	9.9	90	70
Long grafts	0.5–0.8	1.0	118	8.1	95	72

Nanocomposite preparation: The composites were prepared by using two different PCL grades of different molecular weight (PCL40 and PCL80) and silica nanoparticles with three different graft lengths (short, medium, long) and two different target densities (normal and half). Reference PCL samples and samples with unmodified silica particles were also prepared. The composites prepared with the PCL40 matrix were made by first dispersing the required amount of silica nanoparticles to obtain a final concentration of 1, 3 and 5 wt% of silica and 7 g of polymer in toluene. This solution was then solvent casted, left to evaporate in a fume hood overnight and subsequently dried in a vacuum oven at 40 °C overnight. Special care was taken to ensure that no residual solvent was present in the nanocomposite films after drying [25]. The dried samples were then melt-compounded in a recirculating Xplore 15MC (Sittard, Netherlands) micro-compounder at 100 °C, using a mixing time of 10–15 min and a screw speed of 50–60 rpm. For samples based on the PCL80 matrix only 1 wt% of filler was used, but the procedure was the same.

2.3. *Characterisation Methods*

^{1}H Nuclear magnetic resonance (NMR) spectroscopy was performed on a Varian Mercury 400 MHz NMR (Varian Inc., Palo Alto, CA, USA) in CDCl3 to assess if separation of the particles from the reaction mixture after centrifugation was successful.

Fourier transform infrared spectroscopy (FTIR) analysis was performed on a Varian 3100 FTIR spectrometer (Varian Inc., Palo Alto, CA, USA) equipped with a golden gate diamond ATR setup using a spectral range between 600 and 4000 cm^{-1} and 50 scans per spectrum were co-added.

Thermogravimetric analysis (TGA) was performed on a Q500 from TA Instruments Inc. (New Castle, DE, USA) to quantify the amount of grafted material on the modified silica nanoparticles. For this the samples were heated at 10 °C/min from 30–100 °C, kept isothermally at 100 °C for 30 min and then further heated to 800 °C. All the measurements were performed using nitrogen gas (N_2) as a purge flow.

Scanning electron microscopy (SEM) imaging was performed on samples which were dried and cut at −100 °C using a glass-knife and a Leica Ultracut S/FCS microtome (Leica Mikrosysteme GmbH, Wetzlar, Germany). Samples were glued onto a SEM stub and coated with a thin gold layer using an

Emitech K575X sputter coater (Emitech SA, Montigny-le-Bretonneux, France). Samples were thereafter examined in a SEM model XL30 FEG from FEI Company (Eindhoven, Netherlands).

Size exclusion chromatography (SEC) was performed on solutions of 1 mg/mL polymer using tetrahydrofuran (THF) as a solvent. Molecular weight distributions both before and after drying were determined using a SEC from Waters (Elstree, UK) with a Waters 510 pump and a Waters 712 WISP chromatograph with an injection volume of 50 μL. The column used was a PL-gel mix D column from Polymer Laboratories Ltd (Church Stretton, UK). Molecular weights were calculated relative to PS standards and are, thus, not absolute values for PCL.

Polarized optical microscopy (POM) was used to study isothermal crystallisation behaviour and involved the use of a Zeiss Axioplan 2 optical microscope (Carl Zeiss AG, Oberkochen, Germany) equipped with a Zeiss Axiocam camera and polarisers. Thin films of nanocomposite samples placed between two glass slides were prepared with a Tribotrek. During imaging the samples were fixed in a Linkam THMS-600 hot-stage for temperature control (Linkam Scientific Instruments Ltd, Epsom, UK).

Transmission electron microscopy (TEM) was used to study the state of dispersion of the silica particles in the polymer matrix. Ultrathin sections (70 nm) were microtomed at −100 °C using a Leica Ultracut S/FCS microtome. These sections were placed on a 200 mesh copper grid with a carbon support layer. The sections were subsequently examined in a FEI Tecnai 20 transmission electron microscope (FEI Company, Eindhoven, Netherlands), operated at 200 kV.

Differential scanning calorimetry (DSC) using a Q1000 DSC from TA Instruments Inc. (New Castle, USA) was used to study the thermal behaviour of the PCL/silica nanocomposites. Standard aluminium pans were used and a typical sample weight was 3–5 mg. The samples were first heated to 100 °C and equilibrated for 5 min in order to remove thermal history. Subsequently the samples were cooled at 10 °C/min to −80 °C where they were kept for 10 min. Subsequently, they were reheated to 100 °C at 10 °C/min and kept isothermally for 5 min and then cooled again to −20 °C at 10 °C/min. For the isothermal crystallisation experiments, the samples were cooled at 10 °C/min to the desired temperature and kept isothermally for at least 30 min (or 90 min at the highest temperatures). Avrami analyses were performed for all measurements and PCL crystallinity was determined using a heat of fusion at 100% crystallinity of 136 J/g [26].

Rheometry was performed using a stress-controlled AR-G2 rheometer from TA Instruments Inc. (New Castle, USA) under nitrogen atmosphere. Different methods were used for the PCL40 and PCL80 samples. The PCL40 based samples were measured using a 25 mm plate-plate geometry and frequency sweeps were performed at 60 and 100 °C in an angular frequency range of 0.1–100 rad/s with a constant strain of 10%. Isothermal crystallisation experiments were conducted using the rheometer for the PCL40 composites. This was done using a Peltier plate assembly with a 20 mm top plate. The samples were first heated to 80 °C for 5 min and then cooled at 30 °C/min to 47 °C. Two measurements were performed for each composite, one time sweep directly upon cooling and one time sweep after applying a shear of 30 s^{-1} to the cooled sample for three seconds. Time sweeps were performed with a deformation of 1% and an angular frequency of 1 rad/s.

3. Results and Discussion

In the first part of this section the outcome from the silanisation and grafting reactions is presented followed by a discussion on the properties of the prepared nanocomposites and the effect of graft length.

3.1. Grafting Reactions

In Figure 2 SEM images of dried silica, silanised silica and grafted silica nanoparticles are shown. From these images it can be seen that the silanisation was successful. The dried silica particles tend to agglomerate very strongly into solid-like structures, but with the silane attached to the surface, this strong agglomeration is prevented and separate nanoparticles are observed. After grafting, the particles show a similar particle-like morphology, although a tendency to agglomerate into larger clusters than in the case of silanised nanoparticles can be observed.

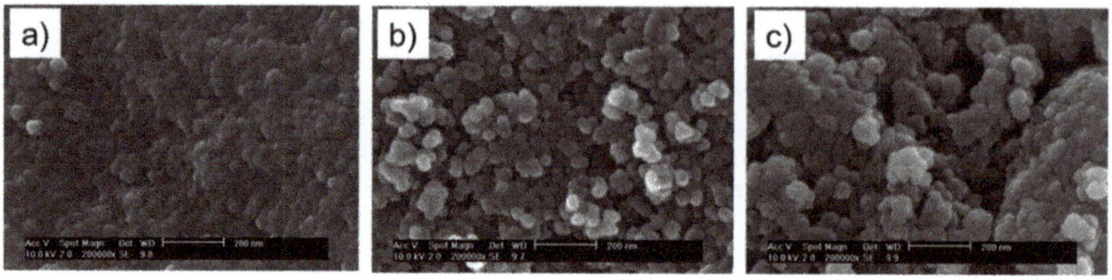

Figure 2. SEM images of: (**a**) silica, (**b**) silanised silica, and (**c**) grafted silica nanoparticles.

The amount of silane and grafted polymer on the particles was estimated from TGA data. The difference in weight loss between unreacted and silanised silica particles was always around 3–4 wt%, which corresponds to a density of about 0.5–0.8 silanes per square nanometre. The TGA results for the grafted particles are shown in Figure 3. From this data the relative weight of the grafts attached to the filler surface was estimated, see Table 2. Since the amount of polymer on the surface increased with reaction time when the other reaction parameters were kept constant, it was assumed that the graft length increased with increasing reaction time. In a similar way, a reduction of the initiator-to-catalyst ratio, while keeping the other parameters constant, results in fewer polymer chain ends attached to the surface. Hence, the grafting density was reduced. Unfortunately, the length of the grafts could not be determined in a quantitative way, but the TGA experiments showed that an increased reaction time lead to increased weight of grafted material for a constant number of initiating sites, an indirect proof that graft length was indeed increased. We will further refer to them simply as short, medium and long grafts.

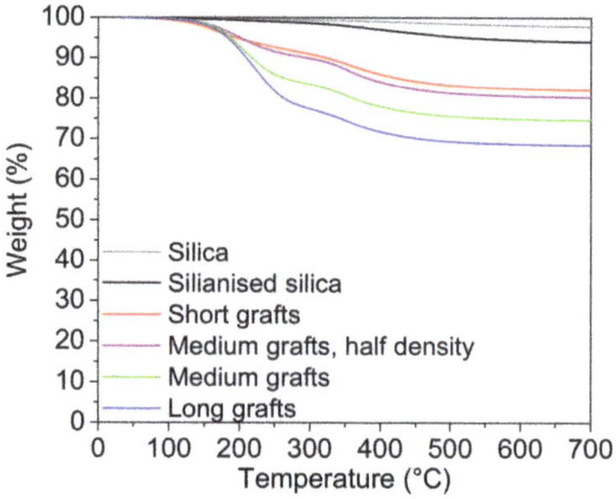

Figure 3. Result from the TGA experiments. Weight is plotted as a function of temperature for untreated silica, silanised silica and different types of grafted silica nanoparticles (short, medium, and long grafts).

Table 2. Weight of the grafted polymers as estimated from TGA data.

Particles	Silanized Particles	Short Grafts	Medium Grafts, Half Density	Medium Grafts	Long Grafts
Est. amount of grafts (% of silica particle weight)	3.8	14.0	15.9	21.5	27.7

The DSC traces for the cooling runs from the melt and the 2nd heating runs of two different grafted particles are presented in Figure 4. Upon cooling, a crystallisation peak is observed for the long grafted particles and upon heating the corresponding melting occurs. Particles with short grafts showed neither a melting nor a crystallisation peak. In order to fold into crystalline lamellae, the polymer chains need to have a certain minimal chain length. From the DSC experiments it is evident that the short grafts are not long enough to crystallise, while the long grafts are sufficiently long enough to crystallize. It has to be noted, however, that both the crystallisation and melting temperatures are lower than the crystallisation and melting temperatures of the neat PCL matrix, which is normally situated around 35–55 °C. This indicates that crystallisation is not favourable and that the formed crystals are not very stable. However, considering these results, it can be expected that the longer grafts will be favourable for nucleation during crystallisation of the nanocomposites.

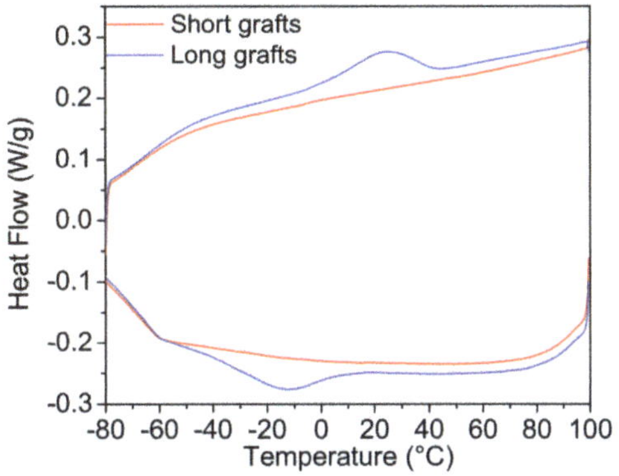

Figure 4. DSC traces of the cooling and the 2nd heating run for silica nanoparticles with short and long graft lengths.

3.2. Dispersion and Morphology

The grafted silica particles were re-dispersed in toluene and mixed with PCL's of two different molecular weights (40 and 80 kg/mol). In addition the samples were also extruded in the micro-compounder to improve the homogeneity of the dispersion. TEM pictures of the obtained morphologies are shown in Figure 5. Morphologies of 3 wt% silica in PCL40 (top row) and 1 wt% silica in PCL80 matrix (bottom row) are depicted. It can immediately be observed that the dispersion of the unmodified silica is very similar for both polymer matrices. Most of the silica nanoparticles are well dispersed in the matrices and very few agglomerates, if any, can be seen. However, dispersion is very different for the grafted silica particles. In both matrices agglomerates are found, although they are generally smaller in the PCL80 system as compared to the PCL40 matrix. There can be various explanations for this observation. One could be attributed to the higher viscosity of the PCL80 system and the development of higher shear forces during the extrusion compounding, leading to better dispersions, in other words, a kinetically driven process. The observed dispersion can also be discussed in terms of relative graft lengths, i.e., a thermodynamically governed process. Akcora et al. [9] showed, both by simulation and experiments, that PS-grafted silica particles in a PS matrix self-ensemble into different structures depending on the relative graft length and graft density. Keeping the grafting density constant while increasing the grafted chain length the particles would first form spherical aggregates followed by more sheet-like and thereafter string-like structures. Only when the graft length was high enough, the particles would be well dispersed throughout the matrix. In a more recent paper on a similar system, Chevigny et al. [12] showed that at a fixed grafting density a crossover ratio exists, where nanoparticle dispersion changes from an agglomerated to a dispersed state. In their case this crossover point could be found at a graft-to-matrix chain length ratio of around 1:4. One obvious difference between our PCL based systems and their systems is the hydrophobic nature of the PS matrix. Due to the hydroxyl groups present on the silica surface, the silica nanoparticles have a strong tendency to agglomerate in hydrophobic matrices. Therefore, well-dispersed nanoparticles could only be obtained in the PS matrix if the molecular weight of the grafted chains is much higher than the molecular weight of the matrix, i.e., a so-called wet brush regime is required. In the PCL matrix, the non-grafted particles are better dispersed than all the grafted particles due to the more hydrophilic nature of PCL. Moreover, considering the DSC results, it is reasonable to assume that the grafted chains in our work are of much lower molar mass than both PCL matrices. Therefore, a certain level of agglomeration can be expected. Increasing the graft lengths did not lead to the formation of different microstructures, since all the graft lengths are relatively short compared to the matrix molar mass. Increasing the molecular weight of the matrix, however, does have the same effect as decreasing the relative grafting length. Therefore, two different morphologies are observed. In the PCL80 matrix small spherical agglomerates can be found, while in the PCL40 matrix the agglomerates are larger and more sheet-like. It must also be mentioned that in all grafted samples some very large agglomerates can be observed. These are formed during the drying step of the grafted nanoparticles, which was necessary in order to analyse them. In the subsequent extrusion compounding step, the shear forces were not high enough to break up all these large agglomerates and, therefore, some of them where still present in the prepared composites. A more ideal way of preparing the nanocomposites might be to add the matrix polymer to the grafted particles in solution directly after polymerization and subsequently evaporate the solvent. The disadvantage of this method is that less information on the grafting reaction can be obtained and, therefore, it was chosen to dry the grafted silica particles prior to composite preparation. The similar state of dispersion of different composites made with the same PCL matrix allows us to study the effect of graft length on crystallisation behaviour in a more straightforward way. Thus, changes in crystallisation behaviour can be directly related to the graft length and not to morphological changes induced by differences in graft lengths.

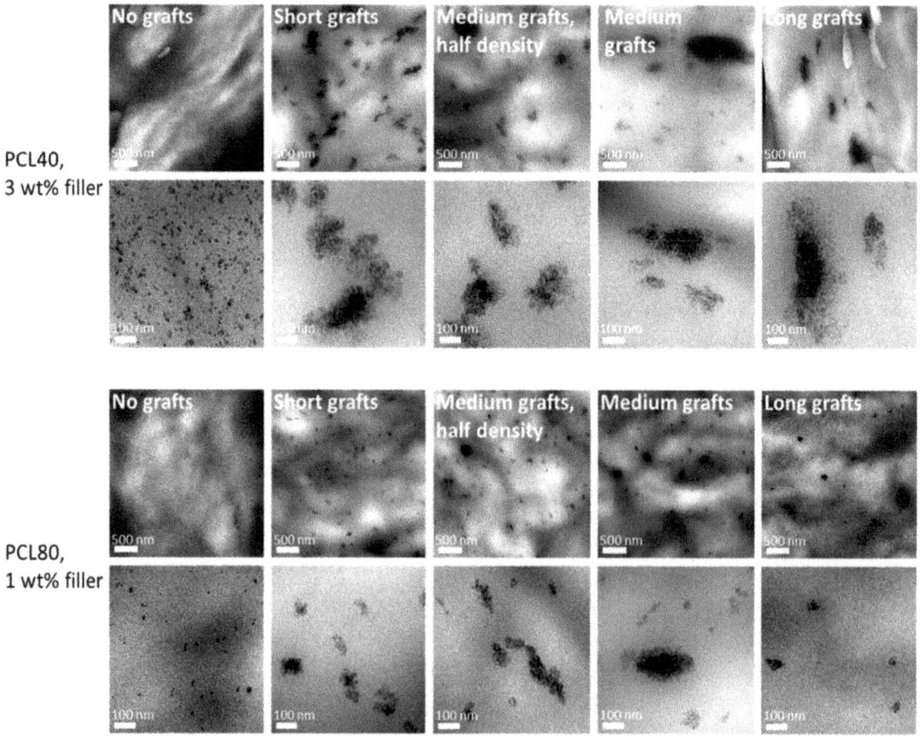

Figure 5. TEM images of PCL/silica composites based on PCL40 matrix and 3 wt% filler (top two rows), and PCL80 matrix and 1 wt% filler (bottom two rows) at two different magnifications. For each top row the scale bar is 500 nm and for each bottom row it is 100 nm.

3.3. Crystallisation Behaviour

3.3.1. Non-Isothermal Crystallization

The crystallisation behaviour of the different PCL/silica composites was investigated using DSC and polarised optical microscopy (POM). In Figure 6 typical cooling traces from DSC experiments are presented. All PCL40 samples show a single crystallisation peak with a maximum around 32 °C. From the value of the onset of crystallization, the nucleation efficiency of the filler can be calculated according to the method first proposed by Fillon et al. for the crystallisation of polypropylene (PP) [27,28]. The maximum temperature for nucleation was determined from self-nucleating experiments and was found to be 46.3 and 45.0 °C, respectively. The crystallisation parameters extracted from the DSC traces can be found together with the nucleation efficiency (N.E.) in Table 3.

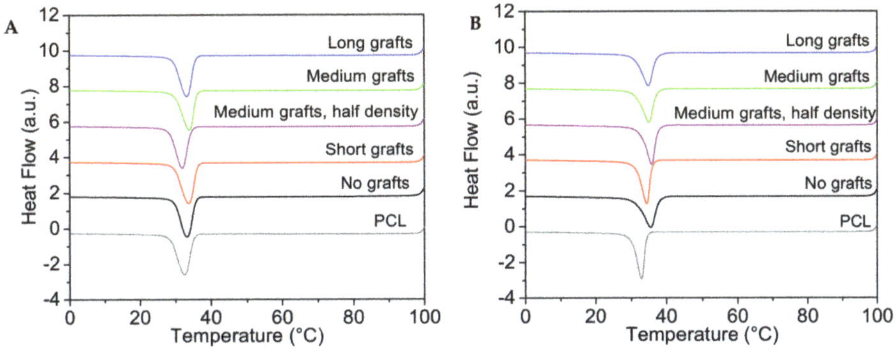

Figure 6. DSC cooling traces for PCL/silica composites based on PCL40 matrix and 1 wt% filler (**A**) PCL/silica composites based on PCL80 matrix and 1 wt% filler (**B**).

The crystallization behaviour of the PCL40 matrix is rather unaffected by the addition of small amounts of untreated nanoparticles. However, the addition of 5 wt% unmodified silica particles to the PCL40 matrix gives a negative nucleation efficiency. This is in line with our earlier data [29] where it was shown that the addition of silica nanoparticles to the same PCL40 matrix hindered crystallisation in continuous cooling experiments. Adding particles with short grafts yields a small nucleation effect, while the addition of nanoparticles with longer grafts is again disadvantageous. Reducing the grafting density is reducing the nucleation efficiency further. This result is in line with the result of L'Abee et al. [24], who showed that in a rubber particle-filled PCL based system, the nucleation efficiency depended on grafting density. For the PCL80 matrix the behaviour is different. The addition of even a small amount of 1 wt% silica nanofiller has a significant nucleating effect on the PCL matrix, evidenced by a shift in the crystallization peaks to higher temperatures. There is a clear effect of grafting length, where shorter grafts in this case are giving lower nucleation efficiency. Interestingly, this is also in line with previous data from our group for another composite system, where cellulose nanocrystals were grafted with either short or long grafts [30]. It was shown that short grafts had a better nucleation ability in the lower molecular weight matrix while longer grafts were rendering a higher nucleation efficiency in the higher molecular weight matrix. The highest nucleation efficiency is found in the sample with non-grafted particles. This can likely be attributed to the better dispersion in this sample. Moreover, a clear influence of graft length can be seen, i.e., the longer the grafts, the better the nucleation efficiency. The data also shows that a decrease in graft density is beneficial for the nucleation efficiency. The melting temperature and degree of crystallinity are rather unaffected by filler addition and the obtained crystallinity is lower for composites based on the PCL80 matrix.

Table 3. Crystallisation parameters for PCL/silica composites based on PCL40 or PCL80 matrix.

Composite	Silica Content (wt%)	T_m (°C)	T_c (°C)	Crystallinity (%)	N.E. (%)
		PCL40			
PCL40	0	56.3	35.8	47.0	0.0
No grafts	1	56.7	35.8	44.8	−0.3
	3	56.4	36.9	45.4	11.6
	5	56.5	34.4	43.5	−15.4
Short grafts	1	56.6	36.3	47.9	5.5
	3	57.0	36.5	46.2	7.3
	5	56.4	36.6	47.0	8.2
Medium grafts	1	57.2	35.9	43.9	1.3
	3	56.4	35.6	46.7	−2.6
	5	56.4	36.2	46.3	4.1
Medium grafts, half density	1	56.4	34.5	45.8	−14.6
	3	56.7	35.4	45.5	−5.0
	5	56.6	35.1	44.2	−7.7
Long grafts	1	56.3	35.8	47.6	−0.7
	3	56.3	35.4	45.4	−4.2
		PCL80			
PCL80	1	56.3	36.4	40.8	0
No grafts	1	56.5	40.9	40.9	32.1
Short grafts	1	56.1	38.4	41.3	13.6
Medium grafts	1	56.8	41.2	39.5	22.5
Medium grafts, Half density	1	56.1	41.4	40.6	27.0
Long grafts	1	56.4	41.0	40.2	25.7

3.3.2. Isothermal Crystallisation

The kinetics of isothermal crystallisation was studied using DSC and POM. It should be noted that the temperature window where the crystallisation experiments can be performed is rather limited. At too high temperatures, i.e., above 47 °C, the crystallisation takes a very long time and experiments are extremely time consuming. At too low a temperature the crystallisation starts during the cooling to the desired temperature and the determination of crystallisation parameters becomes inaccurate. With the addition of a nucleating agent the lower temperature limit is pushed upwards and the range in which the crystallisation kinetics can be compared becomes very narrow. Therefore, the crystallisation experiments were performed at different temperatures for the different composites and data sets in which the crystallisation started before the isothermal crystallization temperature was reached were not used. Avrami analysis was performed on all data sets obtained from the DSC measurements. In its simplest form the crystallisation can be described by

$$X(t) = 1 - e^{-K(T)\cdot t^n} \quad (1)$$

where the Avrami parameters n and $K(T)$ can be determined from the slope and the intercept of the linear fitting of log(-ln(1-X(t)) versus log(t), respectively. In accordance with literature, only values between 0.3 and 20% of relative crystallinity were used for the determination of this parameter [31] After the values of n and $K(T)$ are obtained, the half-time of crystallisation can be calculated via:

$$\tau_{0.5} = \left(\frac{ln2}{n \cdot K(T)}\right)^{\frac{1}{n}}. \quad (2)$$

The rate of crystallisation can be described as the inverse of the half-time of crystallisation. The rate of crystallisation at different temperatures in nanocomposites based on the PCL40 matrix can be found in Figure 7. At higher crystallisation temperatures, the crystallisation rate is very similar for all samples and very slow. At the lowest temperature, we can see that the addition of non-grafted silica particles initially accelerates the kinetics, but at high filler loading the crystallisation is, instead, slower than for the other samples. These results are coherent with the nucleation results, showing that too much well-dispersed filler particles hinders the crystal growth process. Generally, the crystallization rate is increased with increased amount of silica particles. In accordance with the non-isothermal studies, short and medium grafts have the largest positive effect on crystallisation kinetics.

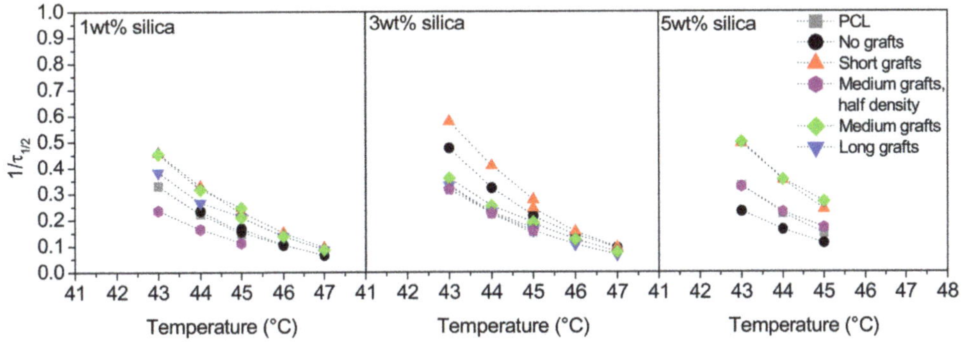

Figure 7. The inverse of half-time of crystallisation versus crystallisation temperature for PCL/silica composites based on PCL40 matrix and 1, 3 and 5 wt% filler (left to right).

In addition to the DSC experiments, isothermal crystallisation experiments were also performed under POM using a Linkam hotstage. These crystallisations were performed at 45 °C for composites based on the PCL40 matrix and the results are depicted in Figure 8. In these images the nucleation ability of the non-grafted silica nanoparticles is clearly visible. Moreover, the dependence of crystallisation kinetics on graft length is clearly shown. Particles with long grafts show a much faster initial crystallisation than their shorter grafted counterparts. At 45 °C, the isothermal DSC results indicated that there is no large difference in crystallization rate between the different grafted particles, thus, these results are somewhat contradictory to the DSC results.

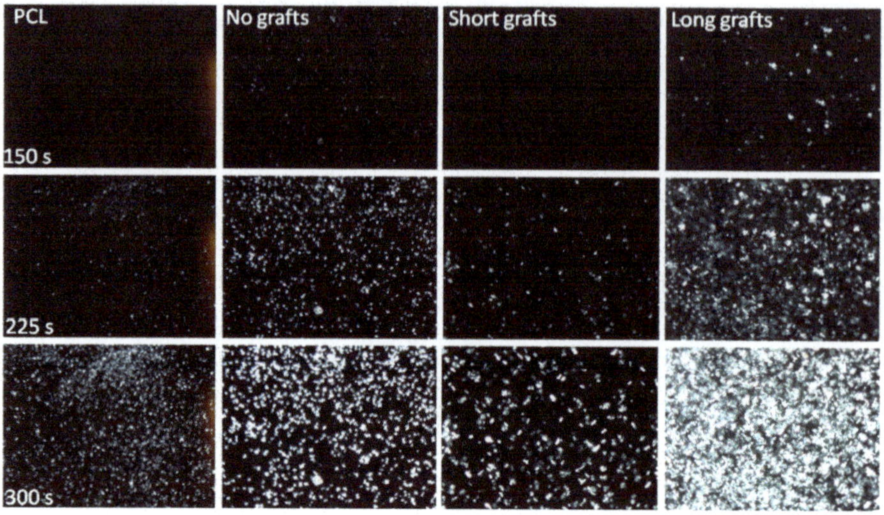

Figure 8. POM images of isothermal crystallisation at 45 °C for different PCL/silica composites based on the PCL40 matrix and 1 wt% filler.

For the PCL80 matrix the results from the isothermal crystallisation experiments are different, but still coherent with the results of the non-isothermal crystallization. The rate of crystallisation versus crystallisation temperature can be found in Figure 9. For all filler types the crystallisation kinetics is accelerated over the whole temperature range with the effect being the largest for non-grafted silica. Shorter grafts are less efficient in accelerating the crystallisation than other fillers, while there is no large difference between medium or long grafts. Moreover, a change in graft density does not have a significant influence.

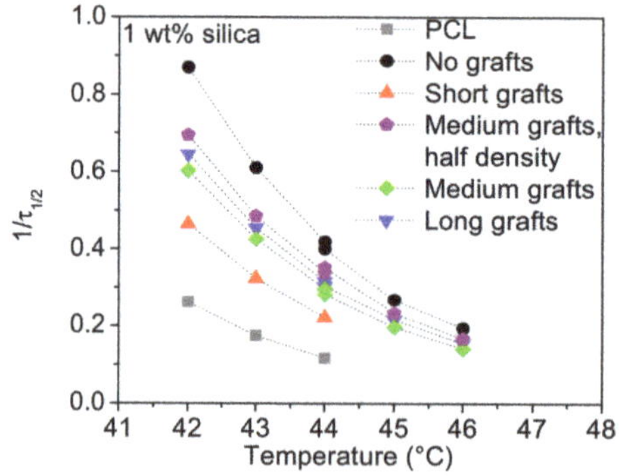

Figure 9. The inverse of half-time of crystallisation versus crystallisation temperature for PCL/silica composites based on the PCL80 matrix.

POM images of samples based on a PCL80 matrix crystallised at 44 °C can be found in Figure 10. These results are in line with the DSC results and confirm the nanofiller's nucleation ability. The influence of graft length is also clearly visible. Longer grafts accelerate the crystallisation to a much larger extent. Interesting to note is that these results are coherent with the results for the PCL40 matrix, even though the DSC results for those samples indicate a slightly different dependence.

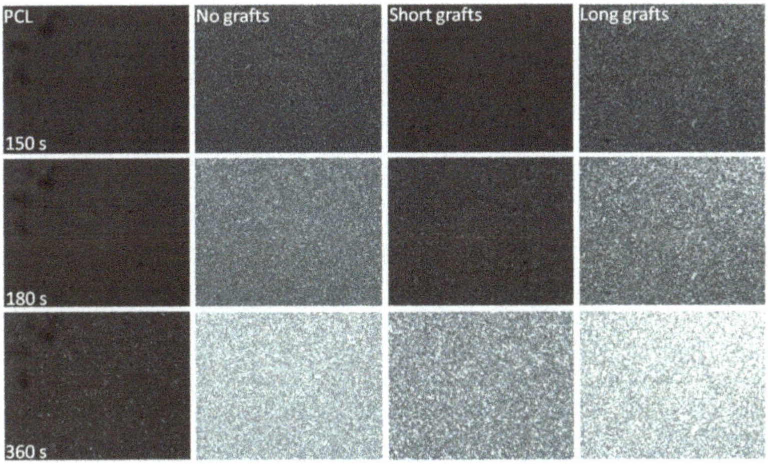

Figure 10. POM images taken during isothermal crystallisation at 44 °C for different PCL/silica composites based on the PCL80 matrix and 1 wt% filler.

From the crystallisation experiments, it can be concluded that the state of dispersion is an important factor governing the crystallisation behaviour of the nanocomposites. Generally, well-dispersed fillers show the highest nucleation efficiency and accelerate the crystallisation to the largest extent. The addition of too much well dispersed nanofiller is, however, not beneficial for both the crystallisation and nucleation efficiency and the overall rate of crystallisation decreases for filler contents exceeding 3 wt%. The grafting of PCL chains to the silica nanoparticles induces agglomeration but to a different extent in the PCL40 and PCL80 matrices. In the PCL40 matrix the large agglomerates yield a very limited nucleation efficiency during continuous cooling experiments, while they seem to accelerate crystallisation during isothermal crystallisation experiments. The influence of graft length is not clearly understood, since some of the employed experimental techniques show different trends. For the PCL80 matrix, the trend is, however, more coherent. Here, grafting of the particles leads to the formation of small silica clusters in the polymer matrix. Even though the nucleation efficiency and the rate of crystallisation is lower for the grafted particles as compared to the non-grafted samples a clear trend can be found in the influence of graft length. Increasing the graft length increases both the nucleation efficiency and the rate of the crystallisation. Decreasing the grafting density does not influence the rate of crystallisation, while the nucleation efficiency decreases.

3.4. *Rheological Behaviour*

3.4.1. Storage, Loss Modulus and Viscosity

The frequency dependence of the storage and loss modulus of the PCL40 based composites are presented in Figures 11–13. The influence of the addition of 1 wt% of silica filler is very marginal for all investigated nanocomposites. At higher filler loading an increase in the storage modulus in the low frequency regime is observed, indicating that the silica nanofillers form a network within the polymer matrix. The loss modulus is less sensitive to filler addition and only at the highest filler content a modest increase is observed.

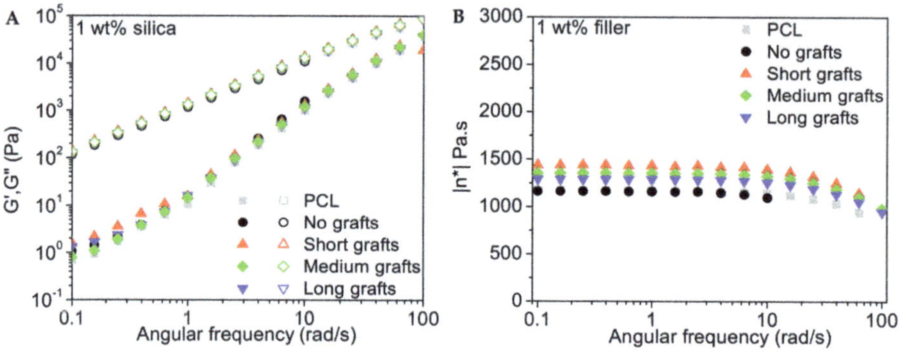

Figure 11. Storage and loss modulus (**A**) and complex viscosity (**B**) as a function of angular frequency for different PCL/silica composites based on the PCL40 matrix and 1 wt% filler. Measurements were performed at 60 °C.

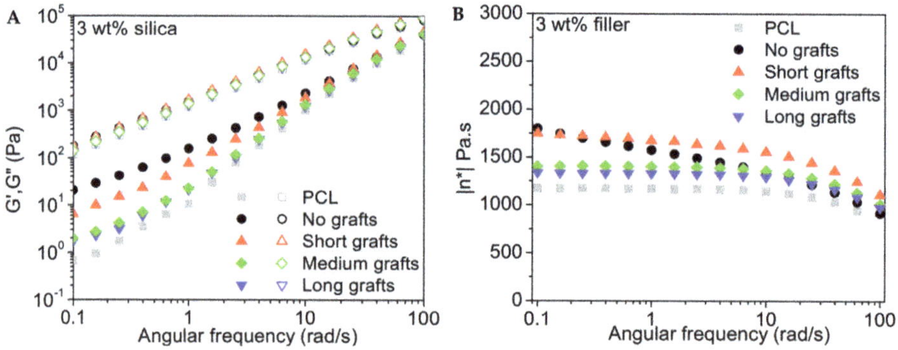

Figure 12. Storage and loss modulus (**A**) and complex viscosity (**B**) as a function of the angular frequency for different PCL/silica composites based on the PCL40 matrix and 3 wt% filler. Measurements were performed at 60 °C.

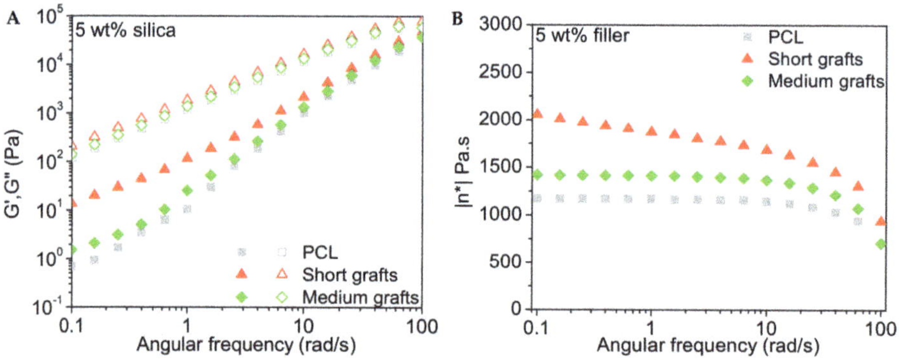

Figure 13. Storage and loss modulus (**A**) and complex viscosity (**B**) as a function of angular frequency for different PCL/silica composites based on the PCL40 matrix and 5 wt% filler. Measurements were performed at 60 °C.

The complex viscosity as a function of angular frequency for the different composites is also presented in the same graphs. It is observed that the relatively low molecular weight of the matrix implies that the viscosity of this material is rather low and quite insensitive to shear thinning in the investigated frequency range. Only at the highest frequencies shear thinning is observed. The addition of 1 wt% silica leads to a minor increase in viscosity but the melt still follows a Newtonian-like behaviour with a close to frequency independent viscosity in the low frequency regime. At higher filler loadings a build-up of the viscosity and a change in slope of the viscosity curve at lower frequencies is observed. This change in slope points to network formation in the melt, which can be attributed to particle-particle interactions. Samples with short graft lengths show the highest increase in viscosity. This might be explained by the difference in molecular weight between the grafts and the bulk matrix. Particles with longer grafts, but still short chain lengths compared to the matrix, contribute to the large amount of low molecular weight PCL in the system, counteracting the effect of the rigid silica fillers. Hence, the effect of filler addition is less pronounced. The data for the PCL80 composites (not shown here) indicates that the addition of 1 wt% filler to the matrix does not alter the properties of the melt significantly, independent of graft length. The viscosity also remained relatively unaffected with only a minor increase in the low frequency regime for the longest grafts.

3.4.2. Shear-Induced Crystallisation

Processing of polymers typically involves flow of a polymer melt followed by a solidification step. We have shown that the addition of various grafted silica nanoparticles affects both the melt properties and the solidification step, in this case crystallisation. Strong flow fields and long relaxation times can promote the crystallisation of oriented structures in the polymer. It is well known that shear can induce crystallisation via the alignment of chains in the direction of shear flow. These aligned chains will form nuclei for the crystallisation and thereby accelerate the crystallization process. If crystallisation takes place under conditions where chain relaxation is slow, row-like nuclei are formed, which promotes the crystallisation of oriented structures, i.e., so-called shish-kebab structures [32]. If the chains are relaxed before the crystallisation sets in, point-like nuclei will be formed. Both types of nuclei accelerate the crystallisation process. The understanding of the influence of fillers on shear-induced crystallisation is not straightforward since several different phenomena are involved. In most cases the addition of rigid fillers leads to an increase in viscosity and decrease in mobility in the polymer melt. This might have advantageous or disadvantageous effects on shear-induced crystallization. On one hand, reduced chain mobility can obstruct the orientation of polymer chains in the direction of shear flow. If the applied shear is not severe enough, fillers remain randomly distributed in the polymer melt and counteract the alignment of the polymer chains, thereby impeding the formation of shear-induced nuclei [30]. However, if the shear flow is sufficiently strong to orient fillers in the direction of flow, they can enhance the nucleation effect through slowing down the relaxation process of the polymer chains after the shear has been removed, with more polymer chains becoming nuclei [33]. In a previous paper we showed that grafted particles affected the polymer melt in a different way depending on graft length [29]. Cellulose nanocrystals with short grafts promoted nucleation with and without shear. However, the grafts hindered the orientation of the filler in the direction of flow, and the resulting crystals where non-oriented spherulites. However, long grafts, but still short in comparison to the bulk matrix, reduced the overall viscosity of the melt and, therefore, also reduced the effect of shear [30].

Here the development of the storage modulus as a function of time at a certain temperature has been used as a measure for crystallisation and the results are generally in good agreement with other measurement techniques, such as DSC [34]. In Figure 14 the development of the storage modulus as a function of time is presented. The experiments were conducted at 47 °C, which was reached via rapid cooling from the melt, and results of both sheared and non-sheared samples are presented. The filler content was 1 wt%. The nucleation effect of the fillers can clearly be seen in both sets of samples, and the results are consistent with the crystallisation data from the DSC measurements. The non-grafted, short chain length grafted, and medium chain length grafted samples show the fastest modulus build up, while fillers with longer grafts are crystallising slower, though faster than the neat PCL matrix. When comparing the sheared with un-sheared samples it should be noted that the effect of adding non-grafted, short grafted and medium grafted particles is greater than the application of a step shear pulse. If shear is applied to these samples, the crystallisation rate is not largely increased since it is already a relatively fast process. This is consistent with results from other nucleated systems in the literature [35]. For nucleated samples, very high shear rates need to be applied in order to further increase nucleation efficiency, since an important number of nucleation sites are already present. The samples with somewhat longer grafts are crystallising slower than samples with fillers without grafts or with short grafts under quiescent conditions and for these samples the shear impulse is an effective way of enhancing nucleation. As shown in the frequency sweep, the viscosity increase is the largest in samples with the shorter grafts indicating that increased fraction of low molecular weight polymer from the longer grafts is counteracting the effect of the addition of hard silica particles. However, since the shear pulse is strong enough to orient the polymer chains and subsequent relaxation of the formed oriented structures is sufficiently slow, a higher nucleation density and a faster overall crystallisation rate as compared to the quiescent condition is observed. The effect of graft density can also clearly be seen. Particles with the lower graft density are crystalizing very slowly under quiescent conditions and the application of a shear pulse is, therefore, greatly accelerating the crystallization. In samples with a higher graft density the crystallization under quiescent conditions is already quite fast and, therefore, there is less influence of the shear pulse.

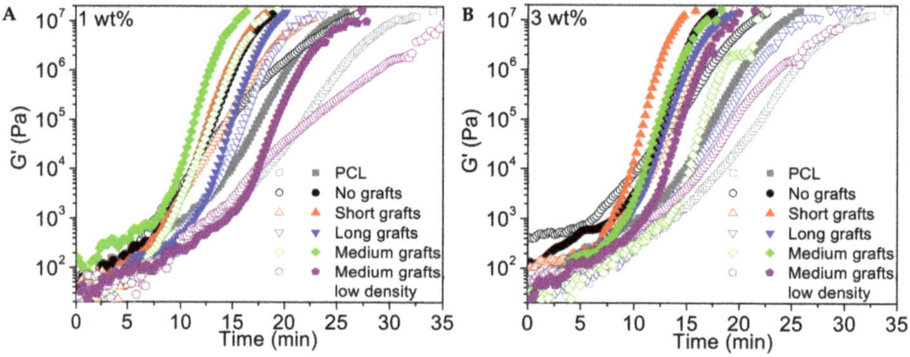

Figure 14. Storage modulus as a function of time for different PCL40/silica composites with 1 wt% filler (**A**) and 3 wt% of filler (**B**). Open symbols corresponds to crystallization experiments without pre-shear and filled symbols to experiments where a pre-shear of 30 s^{-1} was applied for 3 sec. Measurements were performed at a temperature of 47 °C.

Figure 15 shows POM images of PCL40 samples crystallised under different conditions. Clearly, samples subjected to shear show more oriented structures than samples that are crystallised under quiescent conditions. An exception is the neat PCL40 sample, which shows a pure spherulitic morphology also after shearing. The reason for this is that the relaxation time of neat PCL is too short and, therefore, no oriented structures can be formed. The samples with fillers with no or short grafts have longer relaxation times than the neat PCL samples. This allows the samples to retain their orientation after shearing. The samples with the highest viscosity and longest relaxation time during the frequency sweeps were the composites with short grafts, which are also the samples that show the highest orientation after shear.

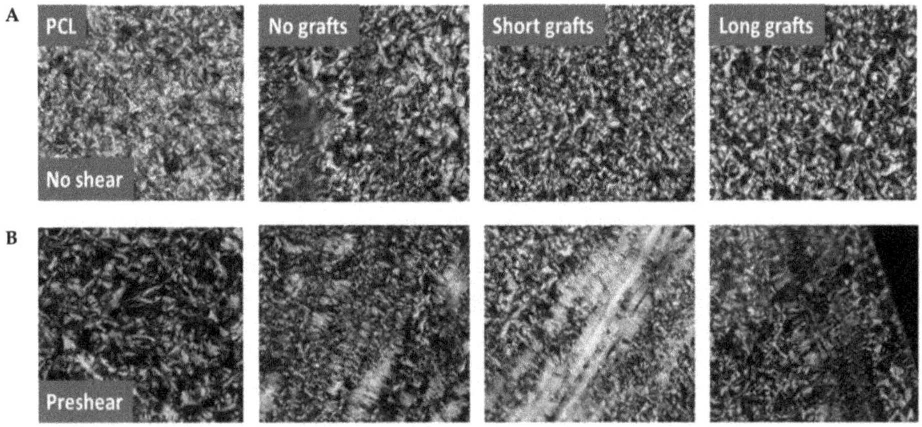

Figure 15. POM images of different samples after crystallisation under different conditions without applied shear (**A**) and with an applied shear of 30 s^{-1} for 3 sec (**B**). Composite samples were based on the PCL40 matrix and contained 1 wt% filler. From left to right: neat PCL, composites with silica particles with no grafts, shorts grafts, long grafts.

4. Conclusions

Silica nanoparticles with different graft lengths have been prepared via a ring opening polymerisation of ε-caprolactone in the presence of silica. The grafted nanoparticles were mixed into PCL matrices with two different molecular weights (PCL40, 40 kg/mol and PCL80, 80 kg/mol) and their morphology and nanofiller dispersion was investigated. It was shown that the ratio between graft length and the molar mass of the matrix is important for the state of dispersion of the nanocomposites. In the lower molecular weight matrix PCL40, no clear influence of graft length on crystallization temperature was found. Compared to neat PCL40, the rate of crystallisation was improved when short or medium graft lengths were used. An increase in filler content led to higher viscosities and a transition to solid-like rheological behaviour in the low frequency regime. The application of a shear pulse increased the rate of crystallization in samples that showed slow crystallization under quiescent condition but was not effective for samples that already showed enhanced crystallization due to the addition of the nanofiller. However, shear promoted the formation of oriented structures. For the higher molecular weight PCL80 matrix the addition of nanosilica acted as an effective nucleation agent, increasing the crystallisation temperature of the sample already at the very low filler content. Increasing the graft length increased the nucleation efficiency without causing any significant changes in the degree of crystallinity or rheological behaviour. However, for all investigated matrices the non-grafted fillers showed the best nucleation ability, which can be attributed to the already excellent dispersion of non-grafted silica in PCL.

Author Contributions: M.E. designed and performed the experiments and analysed the data. J.H. assisted with some of the measurements. H.G. and T.P supervised the project and M.E and T.P wrote the manuscript in consultation with H.G.

Funding: This project was funded by the Dutch Polymer Institute (DPI) under project number #623.

Acknowledgments: The authors greatly acknowledged the Dutch Polymer Institute (DPI) for their financial support.

Conflicts of Interest: The authors declare no conflict of interest.

References

1. Schadler, L.S. Polymer-Based and Polymer-Filled Nanocomposites. In *Nanocomposite Science and Technology*, 1st ed.; Ajayan, P.M., Schadler, L.S., Braun, P.V., Eds.; Wiley-VCH: Weinheim, Germany, 2003; pp. 77–153.

2. Zou, H.; Wu, S.; Shen, J. Polymer/silica nanocomposites: Preparation, characterization, properties, and applications. *Chem. Rev.* **2008**, *108*, 3893–3957. [CrossRef] [PubMed]
3. Tjong, S.C. Structural and mechanical properties of polymer nanocomposites. *Mater. Sci. Eng.* **2006**, *53*, 73–197. [CrossRef]
4. Ou, Y.; Yang, F.; Yu, Z.Z. A new conception on the toughness of nylon 6/silica nanocomposite prepared via in situ polymerization. *J. Polym. Sci. Part B Polym. Phys.* **1998**, *36*, 789–795. [CrossRef]
5. Zhang, R.; Dowden, A.; Deng, H.; Baxendale, M.; Peijs, T. Conductive network formation in the melt of carbon nanotube/thermoplastic polyurethane composite. *Compos. Sci. Technol.* **2009**, *69*, 1499–1504. [CrossRef]
6. Bilotti, E.; Zhang, R.; Deng, H.; Baxendale, M.; Peijs, T. Fabrication and property prediction of conductive and strain sensing TPU/CNT nanocomposite fibres. *J. Mater. Chem.* **2010**, *20*, 9449–9455. [CrossRef]
7. Deng, H.; Skipa, T.; Bilotti, E.; Zhang, R.; Lellinger, D.; Mezzo, L.; Fu, Q.; Alig, I.; Peijs, T. Preparation of high-performance conductive polymer fibers through morphological control of networks formed by nanofillers. *Adv. Funct. Mater.* **2010**, *20*, 1424–1432. [CrossRef]
8. Wang, X.; Foltz, V.J.; Rackaitis, M.; Böhm, G.G.A. Dispersing hairy nanoparticles in polymer melts. *Polymer* **2008**, *49*, 5683–5691. [CrossRef]
9. Akcora, P.; Liu, H.; Kumar, S.K.; Moll, J.; Li, Y.; Benicewicz, B.C.; Schadler, L.S.; Acehan, D.; Panagiotopoulos, A.Z.; Pryamitsyn, V.; et al. Anisotropic self-assembly of spherical polymer-grafted nanoparticles. *Nat. Mater.* **2009**, *8*, 355–359. [CrossRef]
10. Oberdisse, J. Aggregation of colloidal nanoparticles in polymer matrices. *Soft Matter* **2006**, *2*, 29–36. [CrossRef]
11. Akcora, P.; Kumar, S.K.; Garcia Sakai, V.; Li, Y.; Benicewicz, B.C.; Schadler, L.S. Segmental dynamics in PMMA—Grafted nanoparticle composites. *Macromolecules* **2010**, *43*, 8275–8281. [CrossRef]
12. Chevigny, C.; Dalmas, F.; Di Cola, E.; Gigmes, D.; Bertin, D.; Boue, F.; Jestin, J. Polymer-grafted-nanoparticles nanocomposites: Dispersion, grafted chain conformation, and rheological behavior. *Macromolecules* **2011**, *44*, 122–133. [CrossRef]
13. Oh, H.; Green, P.F. Polymer chain dynamics and glass transition in athermal polymer/nanoparticle mixtures. *Nat. Mater.* **2008**, *8*, 139–143. [CrossRef] [PubMed]
14. Wu, T.; Ke, Y. Preparation of silica–PS composite particles and their application in PET. *Eur. Polym. J.* **2006**, *42*, 274–285.
15. Rong, M.Z.; Zhang, M.Q.; Pan, S.L.; Lehmann, B.; Friedrich, K. Analysis of the interfacial interactions in polypropylene/silica nanocomposites. *Polym. Int.* **2004**, *53*, 176–183. [CrossRef]
16. Hasegawa, R.; Aoki, Y.; Doi, M. Optimum graft density for dispersing particles in polymer melts. *Macromolecules* **1996**, *29*, 6656–6662. [CrossRef]
17. McEwan, M.; Green, D. Rheological impacts of particle softness on wetted polymer-grafted silica nanoparticles in polymer melts. *Soft Matter* **2009**, *5*, 1705–1716. [CrossRef]
18. Moon, J.H.; Ramaraj, B.; Lee, S.M.; Yoon, K.R.J. Direct grafting of ε-caprolactone on solid core/mesoporous shell silica spheres by surface-initiated ring-opening polymerization. *J. Appl. Polym. Sci.* **2008**, *107*, 2689–2694. [CrossRef]
19. Joubert, M.; Delaite, C.; Bourgeat-Lami, E.; Dumas, P. Ring-opening polymerization of ε-caprolactone and L—Lactide from silica nanoparticles surface. *J. Polym. Sci. Part A Polym. Chem.* **2004**, *42*, 1976–1984. [CrossRef]
20. Carrot, G.; Rutot-Houzé, D.; Pottier, A.; Degée, P.; Hilborn, J.; Dubois, P. Surface-initiated ring-opening polymerization: A versatile method for nanoparticle ordering. *Macromolecules* **2002**, *35*, 8400–8404. [CrossRef]
21. Vassiliou, A.A.; Papageorgiou, G.Z.; Achilias, D.S.; Bikiaris, D.N. Non-isothermal crystallisation kinetics of in situ prepared poly(ε-caprolactone)/surface-treated SiO_2 nanocomposites. *Macromol. Chem. Phys.* **2007**, *208*, 364–376. [CrossRef]
22. Avella, M.; Bondioli, F.; Cannillo, V.; Di Pace, E.; Errico, M.E.; Ferrari, A.M.; Focher, B.; Malinconico, M. Poly(ε-caprolactone) based nanocomposites: Influence of compatibilization on properties of poly(ε-caprolactone)/silica nanocomposites. *Compos. Sci. Technol.* **2006**, *66*, 886–894. [CrossRef]
23. Zhou, B.; Tong, Z.Z.; Huang, J.; Xu, J.T.; Fan, Z.Q. Isothermal crystallization kinetics of multi-walled carbon nanotubes-graft-poly(ε-caprolactone) with high grafting degrees. *Cryst. Eng. Comm.* **2013**, *15*, 7824–7832. [CrossRef]

24. L'Abee, R.M.A.; van Duin, M.; Goossens, J.G.P. Crystallization kinetics and crystalline morphology of poly(ε-caprolactone) in blends with grafted rubber particles. *J. Polym. Sci. Part B Polym. Phys.* **2011**, *48*, 1438–1448. [CrossRef]
25. Eriksson, M.; Goossens, H.; Peijs, T. Influence of drying procedure on glass transition temperature of PMMA based nanocomposites. *Nanocomposites* **2015**, *1*, 36–45. [CrossRef]
26. Avella, M.; Errico, M.E.; Rimedio, R.; Sadocco, P. Preparation of biodegradable polyesters/high-amylose-starch composites by reactive blending and their characterization. *J. Appl. Polym. Sci.* **2002**, *83*, 1432–1442. [CrossRef]
27. Fillon, B.; Wittmann, J.C.; Lotz, B.; Thierry, A. Self-nucleation and recrystallization of isotactic polypropylene (α phase) investigated by differential scanning calorimetry. *J. Polym. Sci. Part B Polym. Phys.* **1993**, *31*, 1383–1993. [CrossRef]
28. Fillon, B.; Lotz, B.; Thierry, A.; Wittmann, J.C. Self-nucleation and enhanced nucleation of polymers. Definition of a convenient calorimetric "efficiency scale" and evaluation of nucleating additives in isotactic polypropylene (α phase). *J. Polym. Sci. Part B Polym. Phys.* **1993**, *31*, 1395–1405. [CrossRef]
29. Eriksson, M.; Peijs, T.; Goossens, H. The effect of polymer molar mass and silica nanoparticles on the rheological and mechanical properties of poly(ε-caprolactone) nanocomposites. *Nanocomposites* **2018**, *4*, 112–126. [CrossRef]
30. Eriksson, M.; Goffin, A.L.; Dubois, P.; Peijs, T.; Goossens, H. The influence of grafting on flow-induced crystallization and rheological properties of poly(ε-caprolactone)/cellulose nanocrystal nanocomposites. *Nanocomposites* **2018**, *4*, 87–101. [CrossRef]
31. Lorenzo, A.T.; Arnal, M.L.; Albuerne, J.; Muller, A.J. DSC isothermal polymer crystallization kinetics measurements and the use of the Avrami equation to fit the data: Guidelines to avoid common problems. *Polym. Test.* **2007**, *26*, 222–231. [CrossRef]
32. Xu, J.-Z.; Chen, C.; Wang, Y.; Tang, H.; Li, Z.-M.; Hsiao, B.S. Graphene nanosheets and shear flow induced crystallization in isotactic polypropylene nanocomposites. *Macromolecules* **2011**, *44*, 2808–2818. [CrossRef]
33. Chen, Y.H.; Zhong, G.L.; Lei, J.; Li, Z.M.; Hsiao, B.S. In situ synchrotron X-ray scattering study on isotactic polypropylene crystallization under the coexistence of shear flow and carbon nanotubes. *Macromolecules* **2011**, *44*, 8080–8092. [CrossRef]
34. D'Haese, M.; Bart, G.; Van Puyvelde, P. The influence of calcium-stearate-coated calcium carbonate and talc on the quiescent and flow—Induced crystallization of isotactic poly(propylene). *Macromol. Mater. Eng.* **2011**, *296*, 603–616. [CrossRef]
35. Naudy, S.; David, L.; Rochas, C.; Fulchiron, R. Shear induced crystallization of poly (m-xylylene adipamide) with and without nucleating additives. *Polymer* **2007**, *48*, 3273–3285. [CrossRef]

molecules

Article

Polymerization Assisted by Upconversion Nanoparticles under NIR Light

Polina Demina [1,2], Natalya Arkharova [1], Ilya Asharchuk [1], Kirill Khaydukov [1], Denis Karimov [1], Vasilina Rocheva [1], Andrey Nechaev [1,3], Yuriy Grigoriev [1], Alla Generalova [1,2] and Evgeny Khaydukov [1,4,5,*]

1 Federal Scientific Research Center «Crystallography and Photonics» Russian Academy of Sciences, Leninskiy Prospekt 59, Moscow 119333, Russia
2 Shemyakin-Ovchinnikov Institute of Bioorganic Chemistry Russian Academy of Sciences, Miklukho-Maklaya str. 16/10, Moscow 117997, Russia
3 Institute of Fine Chemical Technologies, Moscow Technological University, Vernadsky Avenue 78, Moscow 119454, Russia
4 I.M. Sechenov First Moscow State Medical University, Trubetskaya str. 8-2, Moscow 119991, Russia
5 Institute of Mathematics and Informational Technologies, Volgograd State University, Universitetskiy Prospect, 100, Volgograd 400062, Russia
* Correspondence: khaydukov@mail.ru

Academic Editor: Marinella Striccoli
Received: 3 June 2019; Accepted: 4 July 2019; Published: 5 July 2019

Abstract: Photopolymerization of nanocomposite materials using near infrared light is one of the unique technologies based on the luminescent properties of lanthanide-doped upconversion nanoparticles (UCNPs). We explored the UCNP-triggered radical polymerization both in oligomer bulk and on the nanoparticle surface in aqueous dispersion. Core/shell UCNPs $NaYF_4$:Yb^{3+} and Tm^{3+}/$NaYF_4$ with emitting lines in the ultraviolet and blue regions were used to activate a photoinitiator. The study of the bulk photopolymerization in an initially homogeneous reaction mixture showed the UCNP redistribution due to gradient density occurring in the volume, which led to formation of UCNP superlattices and spheres "frozen" in a polymer matrix. We also developed a strategy of "grafting from" the surface, providing polymer shell growth directly on the nanoparticles. The photosensitization of the endogenous water-soluble photoinitiator riboflavin by the resonance energy transfer from UCNPs was demonstrated in the course of monomer glycidyl methacrylate polymerization followed by photocrosslinking with poly(ethylene glycol) diacrylate on the nanoparticle surface.

Keywords: photopolymerization; nanocomposite materials; upconversion nanoparticles; NIR light; surface modification

1. Introduction

Lanthanide-based upconversion nanoparticles (UCNPs) are a promising nanoplatform for future technologies. UCNPs are a class of photoluminescent material, which have attracted much attention due to their ability to convert low-energy near-infrared (NIR) light into visible and UV photons in the course of the sequential absorption of two or more quanta [1]. UCNPs consist of an inorganic crystalline host matrix, co-doped with a pair of trivalent lanthanide ions (Ln^{3+}), usually with Yb^{3+} (as a sensitizer) and Er^{3+} or Tm^{3+} (as an activator) [2]. NIR light excitation, unlike UV and visible light [3], possesses an ability for deep penetration into biotissues with no photodamage to living cells, leading to the rapid progress of NIR technology in biomedicine. Furthermore, UCNPs are free of the typical drawbacks of conventional fluorophores (quantum dots, nanodiamonds, fluorescence proteins, dyes, etc.) [4,5], and demonstrate high photostability; a lack of blinking and bleaching; narrow emission lines and a large anti-Stokes shift; and low cytotoxicity. All these peculiarities show a highly versatile and translatable UCNP photoluminescent nanotechnology for application in industry and life science [6].

Three-dimensional prototyping of nanocomposite photopolymer materials using NIR light is one of the unique technologies based on the luminescent properties of UCNPs. This process is similar to the process of two photon polymerization (2PP), but at the same time devoid of its shortcomings, such as a long curing process due to the small two-photon absorption cross-sections of used chromophores [7] and, more importantly, the use of a high intensity laser. Unlike the 2PP process, UCNPs can be excited by using semiconductor continuous-wave lasers with relatively moderate intensity, usually 10 W/cm^2 [8]. In addition, this technology inherits all the advantages of 2PP, including the NIR light that falls in the "optical window" with minimal absorption and scattering of typically used monomers and polymers [9]. Under NIR radiation UCNPs can play the role of nanolamps, emitting light deeply inside the composition, which is exploited in order to initiate the photopolymerization of various monomers [10–16] and photocurable materials [17], and to produce 3D objects directly in the bulk of light-sensitive resins [18–20]. Although the NIR photopolymerization strategy is simple and has already been shown, there is still not enough knowledge about the mechanism of the polymerization initiation, polymer chain propagation, distribution of nanoparticles in the composition, and spatial resolution, etc.

NIR light-activated polymerization triggered by UCNPs also gives a new opportunity in the field of nanoparticle surface modification, which is crucial for its biomedical applications. This strategy offers a perspective approach for polymer-shell formation through the chain growth from the UCNP surface. Overlapping the UCNP emission and photoinitiator absorption spectra leads to a polymer shell with controlled thickness attached tightly to UCNP surface, forming a dense layer under NIR irradiation in the dispersion medium, preventing nucleation of new particles. In addition, this approach allows for the embedding of drugs during polymer shell formation on the UCNP surface. To date, only a few works demonstrated NIR-activated surface modification [13,21–23]. Bagheri et al. demonstrated that highly efficient RAFT-polymerization can be used to grow polymer chains directly on the surface of inorganic materials ("grafting from") such as lanthanide doped UCNPs [21]. This polymerization process allowed for production of aqueous nanoparticle dispersions with controlled density and thickness of the polymer shells. Despite all of the advantages (e.g., retaining UCNP optical properties, colloidal stability after polymer shell formation), the RAFT-polymerization is a very complicated process that requires highly-skilled labor and has the potential for hydrolysis of the RAFT-initiator. Xiao et al. demonstrated NIR light-activated photopolymerization of polyethyleneglycol-diacrylate (PEG-DA) on the UCNP $NaYF_4{:}Er^{3+},Yb^{3+}$ surface in the presence of Eosine Y [22]. However, UCNPs tend to aggregate from a uniform dispersion (100–200 nm) to nanoclusters (3–10 mm), while the irradiation time of the NIR light increases. The use of Eosin Y in highly sensitive formulations requires the exclusion of day light or at least the use of special safety light [23]. Beyazit et al. employed oleic acid (OA)-capped UCNPs as an internal light source for photopolymerization in situ of a hydrophilic monomer 2-hydroxyethyl methacrylate (HEMA) and the crosslinker N,N′-ethylenebis- (acrylamide) (EbAM) to obtain a thin polymer shell around the nanoparticles [13]. UCNPs were employed to activate the UV (benzophenone/triethylamine) and visible (eosin Y/triethylamine) initiators and obtain colloidally stable particles of about 50 nm in diameter. However, this technique requires a long period of exposure (4 h). Thus, the development of an approach for a technological polymer shell formation on the UCNPs surface is crucial.

Herein, we report a facile strategy of UCNP-triggered radical polymerization under NIR irradiation both in monomer (oligomer) bulk and on the UCNP surface. We studied bulk polymerization under NIR irradiation in more detail, and showed that, unlike our previous results presented in [19], the utilization of a UCNP concentration below the percolation threshold in a reaction mixture resulted in UCNP redistribution in the reaction volume in the form of superlattices and spheres "frozen" in polymer. We also developed a strategy of "grafting from" the surface, providing polymer shell growth directly on the UCNP under NIR light. The potential of the endogenous water-soluble photoinitiator riboflavin was shown in the course of monomer glycidyl methacrylate (GMA) polymerization followed by photocrosslinking with polyethyleneglycol-diacrylate (PEG-DA) on the surface of UCNPs–GMA. We believe that this quick strategy enables an effective surface modification of nanoparticles, expanding the application of UCNP for industry and biomedicine.

2. Results and Discussion

We used UCNPs β-$NaYF_4$:Yb^{3+}, Tm^{3+} with a core/shell structure, synthesized by a modified solvothermal method, as described in [24]. Nanoparticle size was evaluated by TEM as 40 nm × 30 nm (Figure 1a,b) with a hexagonal (*P6*$_3$*/m* space group) crystal structure favorable for the energy-transfer upconversion process.

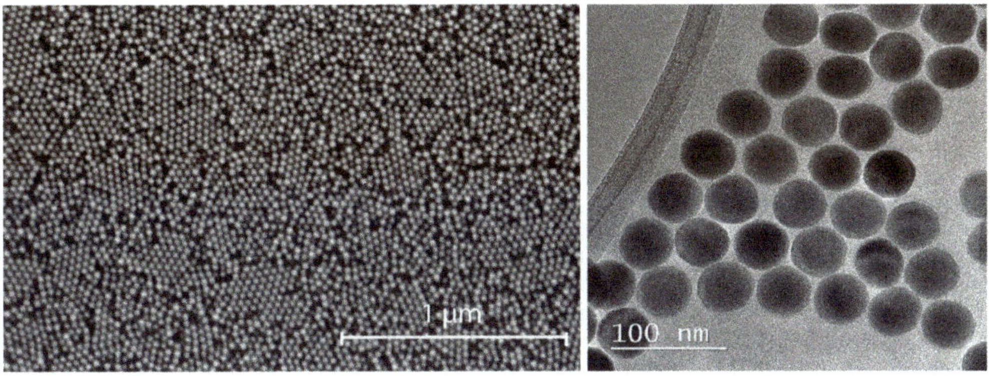

(**a**) (**b**)

Figure 1. (**a**) SEM and (**b**) TEM images of core/shell $NaYF_4$: Yb^{3+}, Tm^{3+}/$NaYF_4$ nanoparticles.

The upconversion mechanism is based on the nonlinear conversion of NIR light through real energy states of trivalent lanthanide ions (see Figure 2a) [25]. Ion Yb^{3+} in the UCNPs absorbs a 975 nm photon and is excited to the $^2F_{5/2}$ state; then, it transfers its energy to the $^3H_5$ state of a neighboring Tm^{3+}. Thereafter, an additional energy transfer occurs from another excited Yb^{3+} to the Tm^{3+}, resulting in further excitation to a higher level of Tm^{3+}. The excited Tm^{3+} is able to emit upconverted photons (ca. 345, 360, 450, 475, 645, or 800 nm) with higher energy levels than those of the exciting photons. As a result, the intensities of the UCNP photoluminescence lines nonlinearly depend on the intensity of the incident NIR light. The general rule can be written as

$$I_{em} \sim [I_{ex}]^n \quad (1)$$

where n is the number of NIR quanta involved in the upconversion process [26].

The photoluminescence spectrum (Figure 2b) of the synthesized UCNPs had strong lines centered on NIR (800 nm), red (645 nm), blue (475, 450 nm), and UV (345, 360 nm) spectral regions under excitation at a wavelength of 975 nm at intensity 10 W/cm^2, which corresponds to 4f transitions of Tm^{3+} ions. Measurements of samples using an integrating sphere showed that nanoparticles had high integral conversion efficiency (~9%) at the excitation intensity 30 W/cm^2.

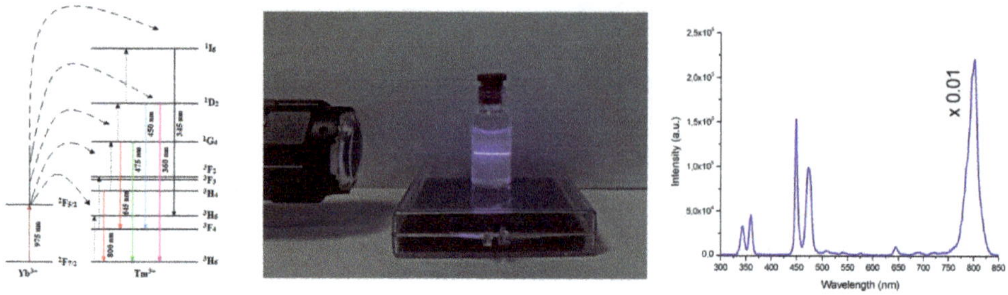

(**a**) (**b**) (c)

Figure 2. (**a**) Energy level diagram of upconversion nanoparticles (UCNPs): $NaYF_4$: Yb^{3+}, Tm^{3+}; (**b**) Visible spectral range image of UCNPs photoluminescence in hexane; (**c**) Corrected apparatus function spectrum of the UCNPs under near-infrared (NIR) excitation at 975 nm at intensity 10 W/cm^2 (line centered at 800 nm multiplied by coefficient 0,01).

2.1. UCNP-Assisted NIR Polymerization in Bulk

In this section, we demonstrate UCNP-assisted photopolymerization of oligocarbonate methacrylate (OCM-2) in the presence of photoinitiator Irgacure 369 under NIR irradiation. In order to start photopolymerization, the initiator should be activated by UCNP light. UCNP wavelengths at 345 and 360 nm fall within the absorption band of the chosen photoinitiator. UCNPs convert near infrared light into UV photons that activate the photoinitiator and start the polymerization process. In our work [19], we explored three-dimensional (3D) rapid prototyping technology based on near-infrared light-induced polymerization of photocurable compositions containing upconversion nanomaterials. We showed that the evaluation of the UCNP concentration threshold in a photocurable composition is required for 3D structure formation due to different polymer growth kinetics on the flat and the narrow edge surfaces of single UCNPs. This study led us to the conclusion that the threshold of UCNP concentration during polymerization should exceed 16.8 mg mL^{-1} in photosensitive compositions for 3D structure formation with high spatial resolution. This phenomenon is explained by the mathematical model of percolation, which has a geometrical-statistical character, for details see [19]. In this work, we focus on the polymerization of photosensitive resin, which contains an order of magnitude lower concentration of UCNPs, namely below the percolation threshold. The rate of UCNP-assisted polymerization non-linearly depends on the NIR light intensity, due to the non-linear nature of the upconversion process. In addition, changing the concentration of nanoparticles in a photocomposition is necessary to maintain the total dose of upconverted UV photons required for the polymerization reaction, which in the first approximation depends linearly on the concentration of nanoparticles in the photocomposition.

Figure 3 shows the SEM-images of a UCNP structure obtained in the course of polymerization. The image on the left side (Figure 3a) demonstrates the structures obtained in the same regime as the structures described in [19] at a nanoparticle concentration of ~ 20 mg mL^{-1}. In the right image (Figure 3b), the composition was photopolymerized at a nanoparticle concentration of ~ 2 mg mL^{-1}. The use of any UCNP concentrations led to the structure formation in NIR irradiated volume. However, with higher magnification (inset), a significant difference in the roughness, despite an initially homogenous chemical photocomposition of the sample, was observed. We believe that this effect is associated with the peculiarities of the photopolymerization process. Since the process of NIR-induced polymerization with UCNPs above the percolation threshold was studied in detail earlier in [19], here we present the structure images in Figure 3a only for comparison of the structure surface morphology. It would be reasonable to assume that UCNP concentration below the threshold results in structures in the form of nanoparticles with a polymer shell. The thickness of the shells may be determined by the mean photon-free path in the photosensitive composition,

$$l_{pl} = \frac{1}{\alpha_{pl}} = \frac{1}{\varepsilon_{pl} \times C_{pl}} \quad (2)$$

where α_{pl}, ε_{pl}, and C_{pl} denote the linear absorption coefficient, molar extinction coefficient, and molar concentration of the photoinitiator, respectively. At a sufficient UV photon flux from the UCNPs and taken that the mean photon-free path in the medium exceeds half of the mean distance between the UCNPs, the polymer shells formed around UCNPs start to interact, leading to a bounded volume. However, in this context, we did not take into account the fluctuation of volumes with different polymer density, which in turn are responsible for the nanoparticle movement during photopolymerization. Nanoparticles are known to diffuse from the illuminated areas into the dark fields under the condition of spatially inhomogeneous illumination of a nanocomposite material. This effect appears as a result of density gradients and, for example, can be used to write phase gratings in a nanocomposite polymer material [25].

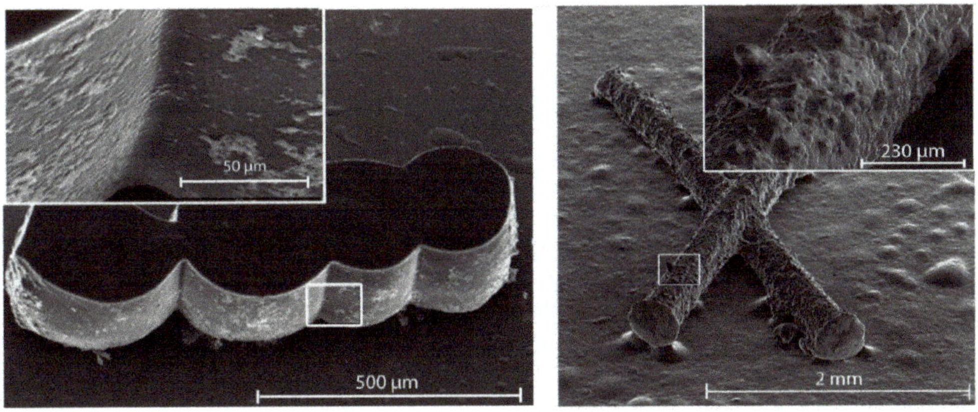

(a) (b)

Figure 3. SEM image of 3D polymer microstructures obtained by NIR light-activated photopolymerization: (**a**) UCNPs concentration ~ 20 mg mL^{-1} and (**b**) UCNPs concentration ~ 2 mg mL^{-1}.

A detailed examination of the structure surface and sections showed that two types of ordered nanoparticle structures distributed randomly can be formed in nanocomposites in the course of photopolymerization: superlattices and spheres (see Figure 4). Formation of superlattices probably occurs due to the frontal polymerization [26] associated with a traveling wave developed under light action in a certain area of photocomposition at the microlevel. Photopolymerization results in a density increase of the photocomposition, providing a density gradient along one of three coordinates. A wave moving along the photocomposition rearranges the embedded nanoparticles to the wave boundary and, at the one-dimensional wave propagation they form tightly packed layers, which sooner or later are "frozen" into the polymer matrix (Figure 4b). Note that another scenario can take place. In the case of a three-dimensional counter-propagating wave (the case of compression), densely packed spherical agglomerates consisting of nanoparticles can be formed (Figure 4a). The analysis of the polymer structure sections made by an ion beam showed such spherical agglomerates of nanoparticles formed by NIR-induced polymerization. It is worth noting that these two processes are random and determined by the stochastic distribution of UCNPs in the composition bulk.

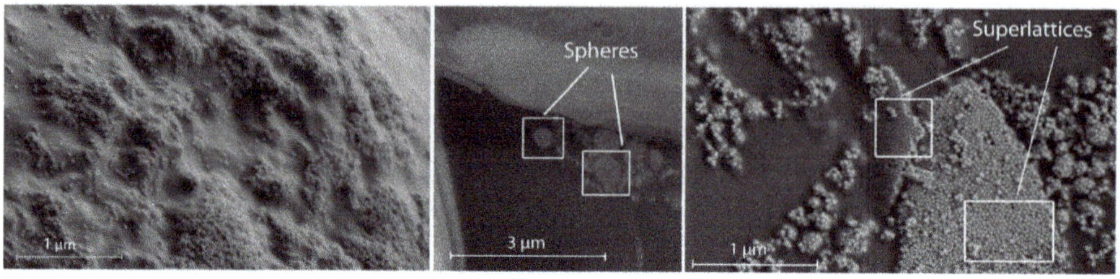

(a) (b) (c)

Figure 4. (**a**) SEM photograph of the polymer structure region with a disordered distribution of UCNPs. Two types of ordered nanoparticle structures formed in the nanocomposite in the course of photopolymerization: spheres (**b**) and superlattices (**c**).

2.2. Graft Surface Polymerization Onto UCNP in Dispersion Under NIR Irradiation

In this section, we consider NIR-induced polymerization in UCNP aqueous dispersion, not in the monomer (oligomer). Photopolymerization under NIR light is one of the challenging methods for UCNP modification and the design of hybrid structures for biomedical applications. Our task was to modify the UCNP surface with polymer shells produced under NIR light. Modification governs the formation of polymer shells comprised of biocompatible material, with controlled thickness, firmly attached to the surface without polymer particle nucleation in the dispersion medium. This approach requires low UCNP concentration and hydrophilization in order to ensure colloidal stability and growth of polymer chains from the surface for initiation of "grafting from" polymerization.

Only a few works addressed NIR-activated photopolymerization in the presence of UCNPs [13,21,22]. Typically, the use of hydrophilic UCNP nanoparticles in vivo, with modification under NIR light as described in the literature, is limited by exploitation of exogenous photoinitiators in the polymerization. There are a large number of photoinitiators characterized by water solubility, biocompatibility, and lack of a cytotoxic effect with 2,2′-azobis[2-methyl-N-(2-hydroxyethyl)propionamide] (VA-086), lithium phenyl-2,4,6-trimethylbenzoylphosphinate (LAP), Irgacure 2959 (1-[4-(2-hydroxyethoxy)-phenyl]-2-hydroxy-2-methyl-1-propane-1-one), Eosin-Y, and others [27]. However, only a few photoinitiators reported in the literature are utilized as complexes with UCNPs irradiated with NIR light (e.g., Eosin-Y, Riboflavin) [28–30]. Here, we applied the endogenous, FDA-approved, non-toxic compound riboflavin in its water-soluble form flavin mononucleotide (FMN) as the photoinitiator for polymerization. Under UV and visible light, FMN was able to produce active radicals in the presence of the coinitiator triethanolamine (TEOHA) [31], which activates polymerization.

We studied the NIR light-activated photopolymerization in the presence of UCNP concentrations under the percolation threshold. Low UCNP concentrations provide thin shell formation on the surface that prevent UCNP aggregation and nucleation of new polymer particles. In general, the polymerization design includes the UCNP surface hydrophilization with tetramethylammonium hydroxide (TMAH); modification with ethylene diamine (EDA), FMN, and glycidyl methacrylate (GMA); polyethylene glycol diacrylate (PEG-DA) immobilization onto the UCNP surface; followed by NIR light triggering for nanoparticle surface modification.

The modification process was carried out in several stages. In order to increase the affinity between the UCNP surface and both the monomer and photoinitiator, an excess of oleic acid (OA) was removed from the UCNP surface with TMAH, then the surface was stabilized by EDA, ensuring amine groups on the UCNP surface. OA partial removal makes the UCNP surface available to FMN molecules. UV and blue photoluminescent lines of UCNPs (Figure 2b) overlap the FMN absorption band (Figure 5b), thus enabling design of the energy transfer donor–acceptor pair UCNP–FMN, relying on resonance energy transfer processes (RET). To ensure an effective RET from the UCNP to the FMN molecule, the distance between donor–acceptor pair should be 3–10 nm as described earlier [24]. FMN coordinates the rare earth metal ions of the UCNPs owing to the presence of the PO_4^- groups (Figure 5a) [32]. After this modification step, a broadband fluorescence signal from 500 to 625 nm, ascribed to the FMN emission under 975 nm excitation, manifested the energy transfer between UCNP and FMN (Figure 5b). The presence of amine groups on the UCNP surface allows for GMA immobilization at the next step and provides tight monomer GMA coupling and propagation of polymer chains from the UCNP surface.

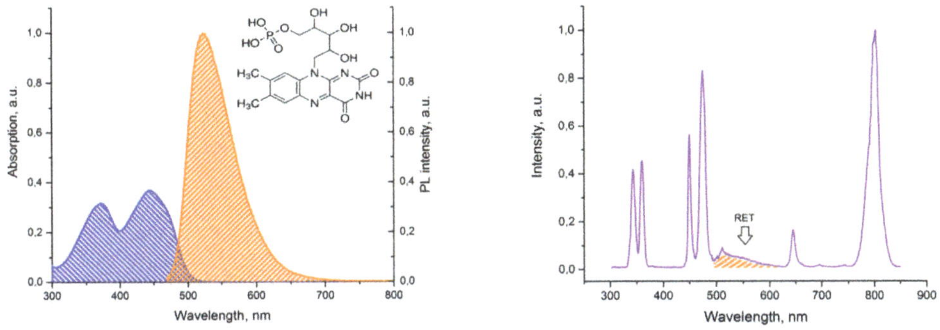

(**a**) (**b**)

Figure 5. (**a**) The excitation (blue) and emission (orange) spectra of flavin mononucleotides (FMN), in insert: FMN formula; (**b**) Photoluminescence of UCNP–FMN in water under 975 nm excitation.

The non-toxic in vivo reagent should possess a biocompatible surface, for example, the same as the PEG molecule can impart. PEG-based materials offer unique biocompatibility and can be used for clinical applications [33,34]. Thus, the next step involved introduction of a hydrophilic biocompatible cross-linker PEG-DA into the system, followed by NIR irradiation to start polymerization GMA. The process was carried out in a scanning mode of irradiation at a laser power of 10 W/cm^2 for 45 min. The UCNP–GMA–PEG sizes were evaluated by dynamic light scattering (DLS) (Figure 6a). The increase of cross-linker concentration led to growth of the UCNP diameters from 120 to 270 nm. At the maximal concentration of PEG-DA (120 μM), the agglomerate formation (740 nm) was observed. TEM images demonstrate that GMA polymerization resulted in a uniform, thin, polymer shell formation (Figure 6b), while the cross-linker addition provided the formation of a knobby polymer shell (Figure 6c,d) with an increase in the uniformity of the shell at higher concentrations of PEG-DA. The different TEM and DLS results can be explained by the hydrophilic nature of polymer coatings that shrink during sample preparation for TEM. In addition, the heterogeneity of the polymer coating that was observed was most likely due to diffusion, insufficient mixing, and the concentration gradient of the polymer inside the composition.

FTIR spectroscopy of the PEG surface-capped particles further confirmed the successful surface modification of the UCNP samples. The results of the comparative analysis of UCNP–TMAH, UCNP–GMA, and UCNP–PEG are presented in Figure 7. A strong stretching vibration of –C=O assigned to oleic acid appeared at 1737 cm^{-1} in the sample treated with TMAH (Figure 7a). However, this characteristic band almost disappeared in the sample of UCNP–GMA before photo-induced crosslinking and in the sample of UCNPs encapsulated in PEG-DA (Figure 7b,c). C–H stretching of the CH_2 polymer groups at 1458 cm^{-1} and 1377 cm^{-1} appear in UCNP–GMA and increase in UCNP–PEG samples. The featured peak at 1654 cm^{-1}, related to the carboxylic groups in UCNP–GMA, disappeared in the UCNP–PEG as a result of the polymerization.

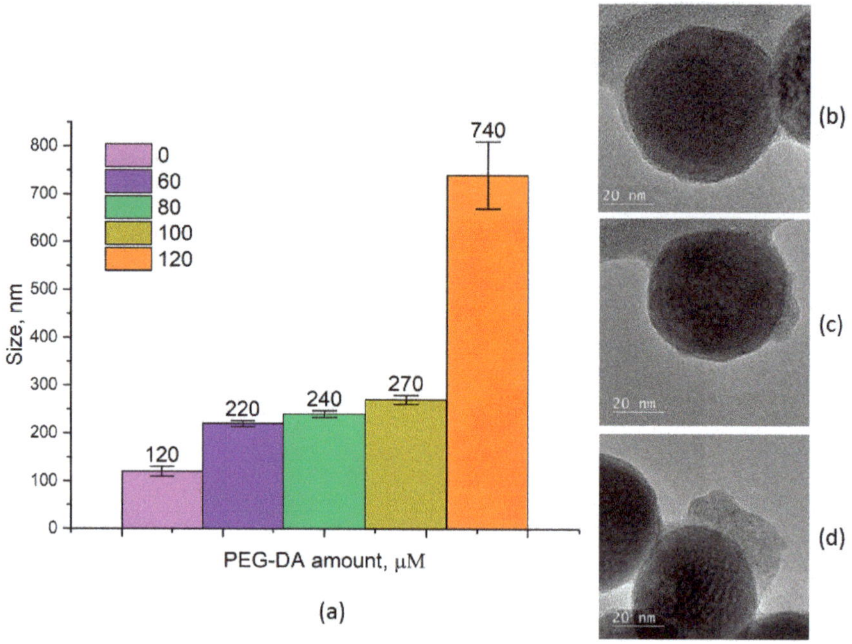

Figure 6. (**a**) Hydrodynamic sizes of UCNP–GMA–PEG via PEG-DA concentration, measured by dynamic light scattering; TEM images of nanoparticles: (**b**) UCNP–GMA; (**c**) UCNP–GMA–PEG, 60 μM PEG-DA; (**d**) UCNP–GMA–PEG, 100 μM PEG-DA. GMA = glycidyl methacrylate; PEG = polyethylene glycol; PEG-DA = polyethylene glycol diacrylate.

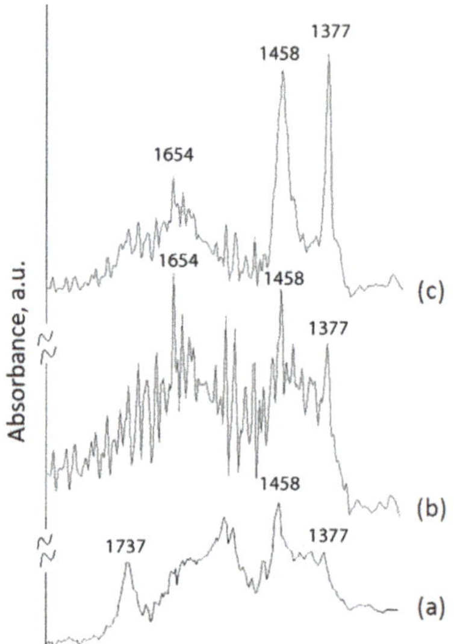

Figure 7. FTIR-spectra of the UCNP samples: (**a**) UCNPs hydrophilized with tetramethylammonium hydroxide (TMAH); (**b**) UCNP–GMA before photo-induced crosslinking; (**c**) UCNPs encapsulated in PEG-DA.

The described method enables photopolymerization followed by cross-linking exclusively on the UCNP surface, eliminating polymer formation in the bulk of the composition and allowing for size control of the UCNP agents. Utilizing non-toxic riboflavin, also called vitamin B_2, as a photoinitiator in combination with a PEG coating allowed for production of a stable aqueous nanoparticle bioprobe. Low radiation intensity, short irradiation time, and adaptability for a specific task are advantages for luminescent probe preparation; these qualites are able to solve optical imaging problems, for example, in passive tumor labeling, where PEG-modified upconversion nanoparticles have proven to be advantageous [35].

The proposed approaches can be adapted for a number of applications based on the delivery of initial liquid materials to the subsurface and confined spaces followed by solidification and shaping. UCNP-based UV emission under NIR light irradiation enables "delivery" of the irradiation to the photocomposition at certain depths. At the same time, the low intensity of the NIR light means this technique has the potential to be employed for various biomedical applications. For example, the first approach can be used for direct writing to prepare the macro- and microstructural configuration of various implants. The second approach can be useful for the UCNP hydrophilization and functionalization required for bioreagents used in biomedical analyzes and visualization.

3. Materials and Methods

3.1. Materials

The following materials were purchased from Sigma-Aldrich (USA), and used without further purification: oligocarbonate methacrylate (OCM-2), photoinitiator Irgacure 369, triethanolamine (TEOHA), dimethyl sulfoxide (DMSO), tetramethylammonium hydroxide (TMAH), ethylene diamine (EDA), glycidyl methacrylate (GMA), polyethylene glycol diacrylate (PEG-DA), Mm ~ 575, potassium bromide, hexane, and chloroform. Flavin mononucleotide (FMN) was purchased from Pharmstandart (Dolgoprudny, Russia).

3.2. Methods

Absorbance characteristics were measured using a Cary 50 (Varian, Palo Alto, CA, USA). Luminescence spectra were recorded by a spectrofuorometer Fluorolog 3 (Horiba Jobin Yvon, France). The conversion efficiency of the nanoparticles was measured by an integrating sphere (Labsphere, North Sutton, NH, USA). FTIR spectra were recorded using an FTIR spectrophotometer (Varian 3100, Palo Alto, CA, USA). The TEM images of the UCNPs were obtained on a transmission electron microscope FEI Osiris (FEI, Hillsboro, OR, USA). Morphology of the nanocomposite polymer samples were investigated by a scanning electron microscope Scios (FEI, Hillsboro, OR, USA) at 1 kV. Cross-sections of the polymer samples were made by a focused Ga+ ion beam technique at 30 kV accelerating voltage. Particle sizes were measured using a particle analyzer T90Plus (Brookhaven Corporation Instruments, Holtsville, NY, USA).

3.2.1. UCNP-Assisted NIR Polymerization in Bulk

A mixure of oligocarbonate methacrylate (OCM-2), photoinitiator Irgacure 369, and nanoparticles was used as the photocurable composition for the UCNP-assisted NIR polymerization in bulk. First, 10 mg of Irgacure 369 was mixed with 1 g OCM-2, then, 100 μL UCNP dispersed in hexane (c = 0.2 g/mL) was added and the mixture was thoroughly shaken and sonicated until the hexane evaporated. The photopolymerization occurred in a laser beam (975 nm) focused into a specific volume of glass vial containing the photocurable composition at a laser power of ~ 100 W/cm^2 for 20 s.

3.2.2. Graft Surface Polymerization of UCNP in Dispersion Under NIR Irradiation

A volume of 40 μL UCNP dispersed (20 mg mL^{-1}) in chloroform was added to 1 mL of a 3% aqueous solution of TMAH, thoroughly shaken, and sonicated until chloroform evaporated.

The obtained probes were purified with centrifugation at 13,400 rpm for 10 min. The supernatant was replaced with distilled water. The obtained probes were stabilized with ethylene diamine (EDA) by incubation in an EDA 0.3% solution for 24 h.

To obtain UCNP–FMN nanocomplexes, UCNP–EDA nanosystems (c=0.8 mg ml^{-1}) were incubated with an FMN solution (c = 10 mg mL^{-1}) for 30 min with constant stirring at room temperature. Non-included FMN were removed by centrifugation at 13,400 rpm for 10 min. The supernatant was replaced with distilled water. The procedure for centrifuging and replacing the supernatant was repeated three times. As a result, hydrophilic UCNP–FMN nanocomplexes with effective energy transfer from nanoparticles to FMN molecules were observed.

The presence of amine groups on the UCNP surface allowed for GMA immobilization at the next step. Aqueous media UCNP–FMN was replaced with dimethyl sulfoxide (DMSO) and 20 μL of GMA was added to the DMSO dispersion of UCN–PFMN and incubated for 24 h. Just ahead of the polymerization process, 60, 80, 100, or 120 μM of cross-linking agent polyethylene glycol diacrylate (PEG-DA) and 2 μL triethanolamine (TEOHA) (c = 1 mg mL^{-1}) were added to the photocurable composition. The polymerization was carried out in a scanning mode of irradiation at a laser power of 20 W/cm^2 for 45 min. The resulting probes were purified with centrifugation at 13,400 rpm for 10 min. As a result, hydrophilic UCNP–PEG nanocomplexes were observed.

3.2.3. Fourier-Transform Infrared (FTIR) Spectroscopy

Pure UCNP–PEG were thoroughly grounded and then pressed with KBr to form a tablet. The samples of UCNP were modified with PEG-DA at 60 and 100 nmol and were further modified with GMA prepared as potassium bromide (KBr) pellets. FTIR spectra were recorded using an FTIR spectrophotometer (Varian 3100, Palo Alto, CA, USA).

4. Conclusions

We have developed two approaches of NIR light-activated polymerization triggered by UCNPs. The first approach based on NIR-induced polymerization of oligomers in bulk in the presence of a typical photoinitiator resulted in the formation of a polymer structure with embedded superlattices and spherical aggregates. This process is associated with two mechanisms of polymerization under NIR irradiation: 1. due to the initiator activation generating radicals as a result of UCNP-induced conversion of NIR light into UV photons; 2. due to the frontal polymerization at a microscopic level as a result of the polymer density gradient. The second approach was based on polymer shell formation on the UCNP surface in an aqueous dispersion medium. The sensitization of the photoinitiator under NIR irradiation by the resonance energy transfer from UCNPs was achieved with the aim of demonstrating the great potential of the biocompatible, non-toxic endogenous reagent FMN for the production of PEG-modified UCNPs that are in high demand for industry and biomedical applications.

Author Contributions: P.D., I.A., K.H., and V.R. performed the experiments and prepared figures and manuscript. A.N. synthesized UCNPs. N.A. and Y.G. performed SEM and TEM images. A.G., D.K., and E.K. analyzed the data and interpreted the experiments. A.G. and E.K. provided the idea and designed the experiments. E.K. coordinated the group.

Funding: This work was supported by the Ministry of Science and Higher Education within the State assignment FSRC «Crystallography and Photonics» RAS in the part of «SEM and TEM images», Russian Science Foundation (Project No. 18-79-10198) in part of «polymerization in bulk», Russian Foundation for Basic Research (Project No 18-29-01021) in the part of «UCNPs modification» and (Project No 18-29-20064) in the part of «structures formed in nanocomposite».

Conflicts of Interest: The authors declare no conflict of interest.

References

1. Haase, M.; Schäfer, H. Upconverting nanoparticles. *Angew. Chem. Int. Ed.* **2011**, *50*, 5808–5829. [CrossRef] [PubMed]

2. Generalova, A.N.; Chichkov, B.N.; Khaydukov, E.V. Multicomponent nanocrystals with anti-Stokes luminescence as contrast agents for modern imaging techniques. *Adv. Colloid Interface Sci.* **2017**, *245*, 1–19. [CrossRef] [PubMed]
3. Yang, D.; Ma, P.; Hou, Z.; Cheng, Z.; Li, C.; Lin, J. Current advances in lanthanide ion (Ln3+)-based upconversion nanomaterials for drug delivery. *Chem. Soc. Rev.* **2015**, *44*, 1416–1448. [CrossRef] [PubMed]
4. Wang, F.; Banerjee, D.; Liu, Y.; Chen, X.; Liu, X. Upconversion nanoparticles in biological labeling, imaging, and therapy. *Analyst* **2010**, *135*, 1839–1854. [CrossRef] [PubMed]
5. Chatterjee, D.K.; Gnanasammandhan, M.K.; Zhang, Y. Small upconverting fluorescent nanoparticles for biomedical applications. *Small* **2010**, *6*, 2781–2795. [CrossRef]
6. Wang, F.; Han, Y.; Lim, C.S.; Lu, Y.; Wang, J.; Xu, J.; Chen, H.; Zhang, C.; Hong, M.; Liu, X. Simultaneous phase and size control of upconversion nanocrystals through lanthanide doping. *Nature* **2010**, *463*, 1061–1065. [CrossRef]
7. Whitby, R.; Ben-Tal, Y.; MacMillan, R.; Janssens, S.; Raymond, S.; Clarke, D.; Jin, J.; Kay, A.; Simpson, M.C. Photoinitiators for two-photon polymerisation: Effect of branching and viscosity on polymerisation thresholds. *RSC Adv.* **2017**, *7*, 13232–13239. [CrossRef]
8. Bagheri, A.; Arandiyan, H.; Boyer, C.; Lim, M. Lanthanide-Doped Upconversion Nanoparticles: Emerging Intelligent Light-Activated Drug Delivery Systems. *Adv. Sci.* **2016**, *3*, 1500437. [CrossRef]
9. Yin, A.; Zhang, Y.; Sun, L.; Yan, C. Colloidal synthesis and blue based multicolor upconversion emissions of size and composition controlled monodisperse hexagonal $NaYF_4$:Yb,Tm nanocrystals. *Nanoscale* **2010**, *2*, 953–959. [CrossRef]
10. Xie, Z.; Deng, X.; Liu, B.; Huang, S.; Ma, P.; Hou, Z.; Cheng, Z.; Lin, J.; Luan, S. Construction of Hierarchical Polymer Brushes on Upconversion Nanoparticles via NIR-Light-Initiated RAFT Polymerization. *ACS Appl. Mater. Interfaces* **2017**, *9*, 30414–30425. [CrossRef]
11. Darani, M.K.; Bastani, S.; Ghahari, M.; Kardar, P.; Mohajerani, E. NIR induced photopolymerization of acrylate-based composite containing upconversion particles as an internal miniaturized UV sources. *Prog. Org. Coat.* **2017**, *104*, 97–103. [CrossRef]
12. Chai, R.; Lian, H.; Cheng, Z.; Zhang, C.; Hou, Z.; Xu, Z.; Lin, J. Preparation and characterization of upconversion luminescent NaYF4:Yb, Er (Tm)/PS bulk transparent nanocomposites through in situ polymerization. *J. Colloid Interface Sci.* **2010**, *345*, 262–268. [CrossRef] [PubMed]
13. Beyazit, S.; Ambrosini, S.; Marchyk, N.; Palo, E.; Kale, V.; Soukka, T.; Tse Sum Bui, B.; Haupt, K. Versatile synthetic strategy for coating upconverting nanoparticles with polymer shells through localized photopolymerization by using the particles as internal light sources. *Angew. Chem. Int. Ed. Engl.* **2014**, *53*, 8919–8923. [CrossRef] [PubMed]
14. Liu, R.; Chen, H.; Li, Z.; Shi, F.; Liu, X. Extremely deep photopolymerization using upconversion particles as internal lamps. *Polym. Chem.* **2016**, *7*, 2457–2463. [CrossRef]
15. Ding, C.; Wang, J.; Zhang, W.; Pan, X.; Zhang, Z.; Zhang, W.; Zhu, J.; Zhu, X. Platform of near-infrared light-induced reversible deactivation radical polymerization: Upconversion nanoparticles as internal light sources. *Polym. Chem.* **2016**, *7*, 7370–7374. [CrossRef]
16. Chen, Z.; Wang, X.; Li, S.; Liu, S.; Miao, H.; Wu, S. Near-Infrared Light Driven Photopolymerization Based On Photon Upconversion. *ChemPhotoChem* **2019**. [CrossRef]
17. Stepuk, A.; Mohn, D.; Grass, R.N.; Zehnder, M.; Krämer, K.W.; Pellé, F.; Ferrier, A.; Stark, W.J. Use of NIR light and upconversion phosphors in light-curable polymers. *Dent. Mater.* **2012**, *28*, 304–311. [CrossRef]
18. Tumbleston, J.R.; Shirvanyants, D.; Ermoshkin, N.; Janusziewicz, R.; Johnson, A.R.; Kelly, D.; Chen, K.; Pinschmidt, R.; Rolland, J.P.; Ermoshkin, A.; et al. Additive manufacturing. Continuous liquid interface production of 3D objects. *Science* **2015**, *347*, 1349–1352. [CrossRef]
19. Rocheva, V.V.; Koroleva, A.V.; Savelyev, A.G.; Khaydukov, K.V.; Generalova, A.N.; Nechaev, A.V.; Guller, A.E.; Semchishen, V.A.; Chichkov, B.N.; Khaydukov, E.V. High-resolution 3D photopolymerization assisted by upconversion nanoparticles for rapid prototyping applications. *Sci. Rep.* **2018**, *8*, 3663. [CrossRef]
20. Méndez-Ramos, J.; Ruiz-Morales, J.C.; Acosta-Mora, P.; Khaidukov, N.M. Infrared-light induced curing of photosensitive resins through photon up-conversion for novel cost-effective luminescent 3D-printing technology. *J. Mater. Chem. C* **2016**, *4*, 801–806. [CrossRef]

21. Bagheri, A.; Arandiyan, H.; Adnan, N.N.M.; Boyer, C.; Lim, M. Controlled Direct Growth of Polymer Shell on Upconversion Nanoparticle Surface via Visible Light Regulated Polymerization. *Macromolecules* **2017**, *50*, 7137–7147. [CrossRef]
22. Xiao, Q.; Ji, Y.; Xiao, Z.; Zhang, Y.; Lin, H.; Wang, Q. Novel multifunctional NaYF 4 :Er 3+,Yb 3+ /PEGDA hybrid microspheres: NIR-light-activated photopolymerization and drug delivery. *Chem. Commun.* **2013**, *49*, 1527–1529. [CrossRef] [PubMed]
23. Chen, Z.; Oprych, D.; Xie, C.; Kutahya, C.; Wu, S.; Strehmel, B. Upconversion-Nanoparticle-Assisted Radical Polymerization at λ =974 nm and the Generation of Acidic Cations. *ChemPhotoChem* **2017**, *1*, 499–503. [CrossRef]
24. Khaydukov, E.V.; Mironova, K.E.; Semchishen, V.A.; Generalova, A.N.; Nechaev, A.V.; Khochenkov, D.A.; Stepanova, E.V.; Lebedev, O.I.; Zvyagin, A.V.; Deyev, S.M.; et al. Riboflavin photoactivation by upconversion nanoparticles for cancer treatment. *Sci. Rep.* **2016**, *6*, 35103. [CrossRef] [PubMed]
25. Nazarov, M.M.; Khaydukov, K.V.; Sokolov, V.I.; Khaydukov, E.V. Laser formation of Bragg gratings in polymer nanocomposite materials. *Quantum Electron.* **2016**, *46*, 29–32. [CrossRef]
26. Ishizu, M.; Aizawa, M.; Nakano, W.; Shishido, A.; Kurata, Y.; Barrett, C.J.; Akamatsu, N.; Hisano, K. Scanning wave photopolymerization enables dye-free alignment patterning of liquid crystals. *Sci. Adv.* **2017**, *3*, e1701610.
27. Pereira, R.F.; Bártolo, P.J. 3D Photo-Fabrication for Tissue Engineering and Drug Delivery. *Engineering* **2015**, *1*, 090–112. [CrossRef]
28. Elisseeff, J.; Anseth, K.; Sims, D.; McIntosh, W.; Randolph, M.; Langer, R. Transdermal photopolymerization for minimally invasive implantation. *Proc. Natl. Acad. Sci. USA* **1999**, *96*, 3104–3107. [CrossRef]
29. Savelyev, A.G.; Bardakova, K.N.; Khaydukov, E.V.; Generalova, A.N.; Popov, V.K.; Chichkov, B.N.; Semchishen, V.A. Flavin mononucleotide photoinitiated cross-linking of hydrogels: Polymer concentration threshold of strengthening. *J. Photochem. Photobiol. A Chem.* **2017**, *341*, 108–114. [CrossRef]
30. Wang, J.; Stanic, S.; Altun, A.A.; Schwentenwein, M.; Dietliker, K.; Jin, L.; Stampfl, J.; Baudis, S.; Liska, R.; Grutzmacher, H. A highly efficient waterborne photoinitiator for visible-light-induced three-dimensional printing of hydrogels. *Chem. Commun.* **2018**, *54*, 920–923. [CrossRef]
31. Ahmad, I.; Iqbal, K.; Sheraz, M.A.; Ahmed, S.; Mirza, T.; Kazi, S.H.; Aminuddin, M. Photoinitiated Polymerization of 2-Hydroxyethyl Methacrylate by Riboflavin/Triethanolamine in Aqueous Solution: A Kinetic Study. *ISRN Pharm.* **2013**, *2013*, 1–7. [CrossRef] [PubMed]
32. Jayapaul, J.; Arns, S.; Lederle, W.; Lammers, T.; Comba, P.; Gätjens, J.; Kiessling, F. Riboflavin carrier protein-targeted fluorescent USPIO for the assessment of vascular metabolism in tumors. *Biomaterials* **2012**, *33*, 8822–8829. [CrossRef] [PubMed]
33. Soman, P.; Fozdar, D.Y.; Lee, J.W.; Phadke, A.; Varghese, S.; Chen, S. A Three-dimensional Polymer Scaffolding Material Exhibiting a Zero Poisson's Ratio. *Soft Matter* **2012**, *8*, 4946–4951. [CrossRef] [PubMed]
34. Tibbitt, M.W.; Anseth, K.S. Hydrogels as extracellular matrix mimics for 3D cell culture. *Biotechnol. Bioeng.* **2009**, *103*, 655–663. [CrossRef] [PubMed]
35. Generalova, A.N.; Rocheva, V.V.; Nechaev, A.V.; Khochenkov, D.A.; Sholina, N.V.; Semchishen, V.A.; Zubov, V.P.; Koroleva, A.V.; Chichkov, B.N.; Khaydukov, E.V. PEG-modified upconversion nanoparticles for in vivo optical imaging of tumors. *RSC Adv.* **2016**, *6*, 30089–30097. [CrossRef]

Sample Availability: Samples of the compounds UCNPs are available from the authors.

molecules

MDPI

Article

Polyolefin/ZnO Composites Prepared by Melt Processing

Alojz Anžlovar [1,*], Mateja Primožič [2], Iztok Švab [3], Maja Leitgeb [2], Željko Knez [2] and Ema Žagar [1,*]

1 Department of Polymer Chemistry and Technology, National Institute of Chemistry, Hajdrihova 19, SI-1000 Ljubljana, Slovenia
2 Faculty of Chemistry and Chemical Engineering, University of Maribor, Smetanova ulica 17, SI-2000 Maribor, Slovenia
3 ISOKON d.o.o., Industrijska cesta 16, SI-3210 Slovenske Konjice, Slovenia
* Correspondence: alojz.anzlovar@ki.si (A.A.); ema.zagar@ki.si (E.Ž.); Tel.: +386-1-4760-204 (A.A.); +386-1-4760-203 (E.Ž.); Fax: +386-1-4760-300 (A.A. & E.Ž.)

Academic Editors: Marinella Striccoli, Roberto Comparelli and Annamaria Panniello
Received: 31 May 2019; Accepted: 28 June 2019; Published: 2 July 2019

check for updates

Abstract: Composites of polyolefin matrices (HDPE and PP) were prepared by melt processing using two commercially available nano ZnO powders (Zinkoxyd aktiv and Zano 20). The mechanical and thermal properties, UV-Vis stability, and antibacterial activity of composites were studied. Tensile testing revealed that both nano ZnO types have no particular effect on the mechanical properties of HDPE composites, while some positive trends are observed for the PP-based composites, but only when Zano 20 was used as a nanofiller. Minimal changes in mechanical properties of composites are supported by an almost unaffected degree of crystallinity of polymer matrix. All polyolefin/ZnO composites exposed to artificial sunlight for 8–10 weeks show more pronounced color change than pure matrices. This effect is more evident for the HDPE than for the PP based composites. Color change also depends on the ZnO concentration and type; composites with Zano 20 show more intense color changes than those prepared with Zinkoxyd aktiv. Results of the antibacterial properties study show very high activity of polyolefin/ZnO composites against *Staphylococcus aureus* regardless of the ZnO surface modification, while antibacterial activity against *Escherichia coli* shows only the composites prepared with unmodified ZnO. This phenomenon is explained by different membrane structure of gram-positive (*S. aureus*) and gram-negative (*E. coli*) bacteria.

Keywords: high density polyethylene; polypropylene; ZnO nanoparticles; composites; antibacterial activity; UV-Vis stability; mechanical and thermal properties

1. Introduction

Polymer nanocomposites with inorganic nanoparticles are propulsive fields of research due to innovative combination of properties, arising from the application of inorganic nanofiller. The introduction of nanofillers into the polymer matrices, in most cases, generates relevant problems in terms of their dispersibility [1]. Since inorganic nanoparticles are mostly hydrophilic, they substantially differ in surface energy from the hydrophobic polymer matrices, causing phase segregation of both composite constituents. Nanoparticles are inclined towards formation of stable agglomerates, which are difficult to break up into individual particles and to disperse them uniformly in the host polymer matrix [2]. Due to a relatively small specific interface surface of agglomerates, they prevent efficient transfer of beneficial properties of the nanofiller, related to its nanoscopic dimension and to the host polymer, resulting in 'nanofilled materials' with properties similar to the traditional microcomposites [3]. In accordance with these considerations, it is clear that controlling the dispersion of nanoparticles throughout the polymer matrix is highly important to fully exploit a potential of polymer nanocomposites. In this respect, the development of effective mixing and dispersion procedure is crucial in nanocomposite preparation [4]. Various approaches have been proposed for the manufacturing of polymer nanocomposites with homogeneously dispersed inorganic nanoparticles [5,6]. Possible solutions are the chemical modification of nanoparticle surface by various silanes to reduce surface energy or in-situ synthesis of nanoparticles inside the host polymer matrix [7,8].

Zinc oxide (ZnO), particularly its nanostructures, has recently attracted significant attention as a highly promising material for a broad range of applications [9–11]. ZnO is a frequently used semiconductor with high UV absorption, interesting electrical and optical properties, which strongly depend on the particle size and shape [12,13]. Besides, ZnO influences the thermal and mechanical properties as well as the chemical and physical stability of polymer matrices [14]. ZnO is known for its UV protecting capability [15–17], but its impact on the catalytic degradation of polymers has been less explored [18,19].

In the middle of the 1990s, it was discovered that ZnO also shows antibacterial activity against some bacterial strains. There are many studies on the preparation of ZnO/polymer nanocomposites with antibacterial activity [20,21]. For effective antibacterial activity, ZnO has to come into direct contact with the microorganisms. In the case of composites prepared with surface modified ZnO, antibacterial activity is compromised at the expense of better matrix stability.

Many publications reported on polyolefin/ZnO (nano)composites which were prepared by various processes. Here, we focus only on the composites prepared by melt processing (extrusion and injection molding). Some authors disclosed enhanced electrical properties (reduced resistivity for few orders of magnitude) [22–24] or UV stability [15–17] when large quantities of nano ZnO (30 wt% and more) were added to the polyolefin (PE or PP) matrices. Some of them reported on significantly improved antibacterial activity of PE or PP/ZnO nanocomposites [7,20,21,25]. Considering the mechanical properties of polyolefin/ZnO (nano)composites, some authors reported on significantly enhanced properties [15,26,27], while others stated only a minimal effect of nano ZnO addition [28–30]. Because interfacial interaction between ZnO and polyolefins is very weak [31], the surface of ZnO is frequently modified, mostly with silanes, in order to increase the compatibility between both composite constituents [27,30]. Nano ZnO can increase the degree of crystallinity of polyolefins [31,32], although many authors also reported its negligible effect [22,29]. Therefore, there are still many aspects of polyolefin/ZnO nanocomposites formation and properties, as well as potential applications that need to be cleared out.

Here, we report on the preparation of the composite materials of polyolefin matrices (HDPE and PP) and various types of commercially available nano ZnO (unmodified and surface modified with stearic acid, triethoxycaprylylsilane, and [3-methacryloxypropyl]trimethoxysilane) by deposition of nano ZnO on the surface of polyolefin granules and subsequent melt processing. Our goal was to introduce certain functionalities into the studied polymer matrices, such as UV absorption and antibacterial activity. Besides, the impact of ZnO addition on the composites' mechanical and thermal properties, as well as UV-Vis stability, was studied.

2. Results and Discussion

2.1. Characterization of ZnO Nanofillers

Samples of commercial zinc oxides (Zinkoxyd aktiv and Zano 20) were characterized before they were applied as the nanofillers. The specific surface area based on the BET method is 42.8 m^2/g for Zinkoxyd aktiv and 25.1 m^2/g for Zano 20, while the average pore dimensions are 16.2 nm and 7.6 nm, respectively. SEM micrographs show well-defined ZnO nanoparticles with particle sizes between 20 and 100 nm for both samples (Figure 1). Zano 20 contains a larger fraction of rodlike ZnO structures (Figure 1b), while the degree of particle agglomeration is higher for the Zinkoxyd aktiv. FTIR spectra of ZnO powders (Figure 2A) show characteristic strong and broad absorption bands between 420 and 450 cm^{-1} due to the two transverse optical stretching modes of ZnO [33]. In the FTIR spectrum of Zinkoxyd aktiv, consisting of ZnO nanoparticles with irregular spherical morphology, only broad absorption band with a maximum at 447 cm^{-1} is observed (Figure 2A), while the FTIR spectrum of Zano 20 shows broad band with two maxima, one at 447 cm^{-1} and the other one at 434 cm^{-1} (Figure 2A), which are characteristic of the rod-like ZnO morphology [34]. Besides, Zinkoxyd aktiv shows additional absorption bands at 1508 and 1400 cm^{-1}, typical for the organic moieties most probably located on the surface of ZnO particles, indicating that Zinkoxyd aktiv is more organophillic than Zano 20. Photoluminescence spectra of both ZnO powders show the near band edge UV emission from 380 to 400 nm and numerous visible light emission peaks at 423, 448, 461, 485, and 529 nm [10,12] (Figure 2B). Differences between the samples in the visible light emission region are rather small, indicating small differences in the quantity and type of intrinsic defects on the surfaces of both ZnO samples [35]. A larger difference was observed in the near band edge peak, which is located at 381.5 nm and 392 nm for the Zano 20 and Zinkoxyd aktiv, respectively. The intensity of this peak is much higher for the Zano 20 and is most probably related to the rod-like ZnO morphology (thick rods) [36]. XRD diffractograms (Figure 2C) show diffraction maxima that are characteristic of the crystalline ZnO with hexagonalwurtzite structure (JCPDS card no. 01-079-0205) at 2θ values: 31.8, 34.5, 36.2, 47.6, 56.6, 62.9, 66.4, 67.9, 69.1, 72.6, and 76.9 [37]. Calculated average crystallite sizes are 16.4 nm for Zinkoxyd aktiv and 42.7 nm for Zano 20, indicating that the latter contains much larger crystallites than the former one. Based on these results, we conclude that the Zano 20 has larger crystallite size, lower specific surface area, and rod-like morphology, while the Zinkoxyd aktiv has irregular spherical morphology, larger specific surface area, and organic layer on the surface.

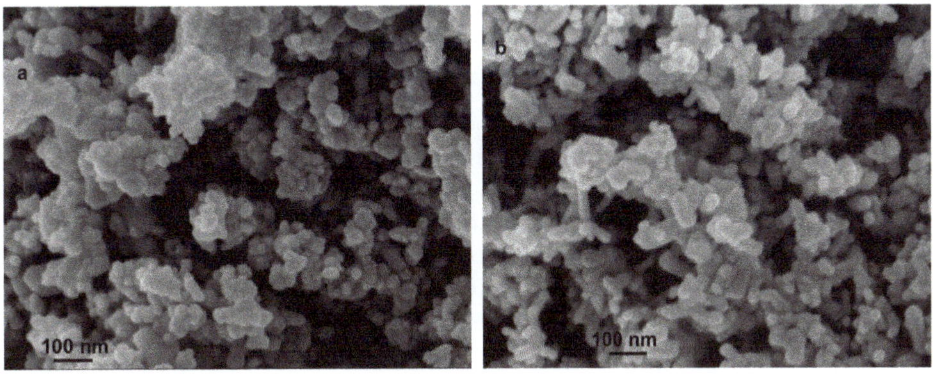

Figure 1. SEM micrographs of (**a**) Zinkoxyd aktiv; (**b**) Zano 20.

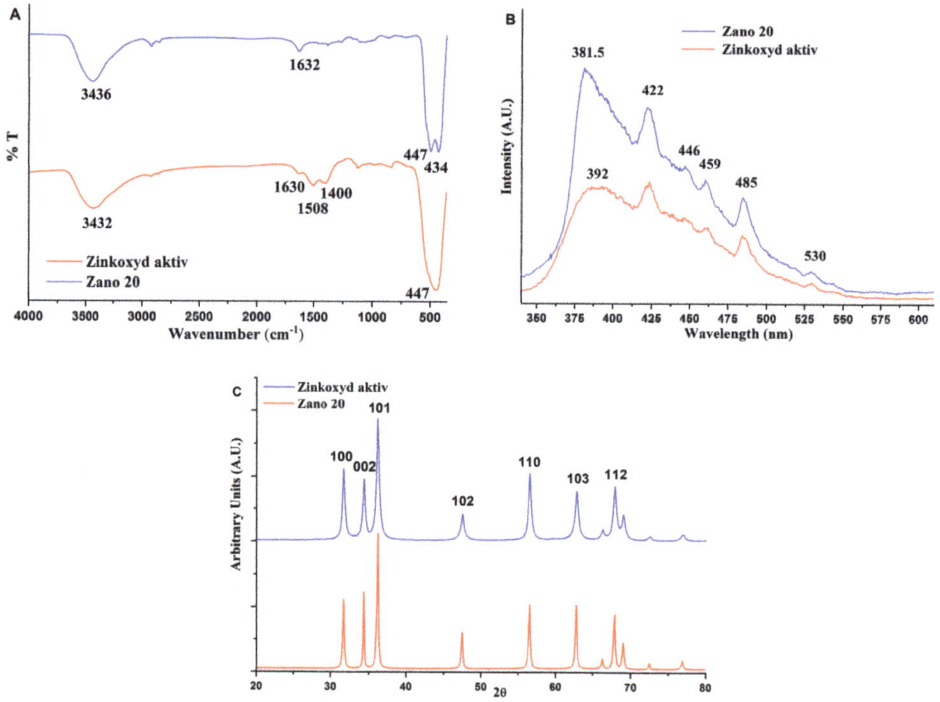

Figure 2. (**A**) FTIR spectra; (**B**) Photoluminescence spectra; (**C**) XRD diffractograms of commercial nano ZnO powders.

2.2. Composites of Nano ZnO with HDPE and PP Matrices; Unmodified ZnO and Surface-Modified ZnO with Stearic Acid

First, we studied the distribution of ZnO in polyolefin matrix by SEM microscopy using a backscattered electron detector. SEM micrographs in Figure S1A–D show the distribution of ZnO in the HDPE matrix at a concentration of nano ZnO of 1.0 wt%. The dimensions of ZnO are from 1 to 5 μm, indicating that ZnO is predominantly in the aggregated form. A comparison of the micrographs of the HDPE nanocomposites prepared by unmodified ZnO (Figure S1A,C) and with stearic acid modified ZnO (Figure S1B,D) indicates that stearic acid coating improves compatibility between the Zano 20 and HDPE, as indicated by a reduced number of large aggregates (Figure S1D).

Table S1 shows the mechanical properties of HDPE/ZnO composites (unmodified ZnO and with stearic acid (3.0 wt%) modified ZnO: Zinkoxyd aktiv and Zano 20) as a function of nano ZnO concentration. The results show that Zinkoxyd aktiv has no particular effect on the composites' mechanical properties since they remain more or less unchanged with a slight downward trend. An exception is elongation at break, which shows a trend of slight increase (Table S1, Figure S2A). The HDPE nanocomposites with Zano 20 show reduction in the tensile strength and Young's modulus and a more pronounced increase in elongation at break (Table S1 and Figure 3). It is obvious that the addition of nano ZnO slightly deteriorates the mechanical properties of the HDPE/ZnO nanocomposites. In particular, the decrease in tensile strength and Young's modulus (by 21% and 25%, respectively) was pronounced when 2% by weight of Zano 20 was added (Figure 3 and Figure S2B). Obviously, such a high concentration of Zano 20 in HDPE is not beneficial. Surface modification of ZnO with stearic acid shows slightly enhanced compatibility between ZnO and HDPE (Figure S1), but no improvement of composites' mechanical properties was observed (Table S1).

Study of thermal properties (melting temperature and melting enthalpy) of HDPE/ZnO composites (unmodified Zinkoxyd aktiv and Zano 20) as a function of nano ZnO concentration shows that the addition of nano ZnO only slightly affects the melting temperature, ΔH_m, and crystallinity degree of HDPE matrix (Table S1, Figure S3). Only small changes in the degree of HDPE crystallinity, together with a rather high degree of ZnO aggregation (Figure S1), are explanations for a minimal impact of added nano ZnO on the nanocomposites' mechanical properties. Due to a high degree of ZnO aggregation, the interface surface between ZnO and HDPE is rather small. On the other hand, a major mechanism influencing the nanocomposite mechanical properties is that inorganic nanostructures act as the crystallization nuclei, resulting in a higher crystallinity of the polymer matrix and thus in improved mechanical properties. Since, in our case, the crystallinity degree is not affected by ZnO, this explains only the small changes in observed mechanical properties (Table S1). In literature, some authors reported on significant increase in crystallinity by the addition of nano ZnO [31], while others observed no changes [22]. Consequently, some authors reported on improved composite mechanical properties [7,15] and others did not [2,28]. On the other hand, TGA results reveal improved thermal stability of the HDPE/ZnO composites compared to the neat HDPE (Figure S4).

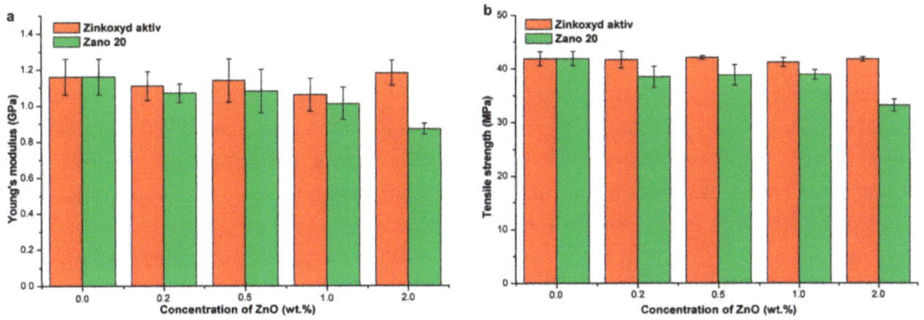

Figure 3. Mechanical properties of HDPE/ZnO composites as a function of nano ZnO concentration: (**a**) Young's modulus and (**b**) Tensile strength.

SEM micrographs in Figure S5A–D show the distribution of ZnO in the PP matrix at 1 wt% concentration of nano ZnO. The ZnO particles are predominantly present in PP matrix as the aggregates with sizes varying from 1 to 5 μm. A comparison of micrographs (unmodified ZnO—Figure S5A,C and ZnO modified with stearic acid, Figure S5B,D) shows that stearic acid primarily improves the compatibility of Zano 20 with PP matrix, as indicated by the smaller number of large aggregates (Figure S5D).

The results of mechanical properties testing of PP/ZnO composites (unmodified ZnO and modified with stearic acid: 3.0 wt%) prepared by Zinkoxyd aktiv or Zano 20 nanofillers as a function of nanoparticle concentration are presented in Table S2 and Figure 4. Zinkoxyd aktiv does not have a significant influence on the composites' mechanical properties at concentrations up to 2.0 wt% (Table S2, Figure S6A). On the other hand, Zano 20 shows a slightly more pronounced effect as indicated by increased tensile strength by 7.3% and Young's modulus by 6.3% (Table S2, Figure S6B). Additionally, unmodified ZnO slightly increases Young's modulus, while this was not observed for the ZnO modified with stearic acid, which is attributed to the plasticizing effect of stearic acid.

Table S2 also summarizes the thermal properties (melting temperature and melting enthalpy) of PP/ZnO composites (Zinkoxyd aktiv and Zano 20) depending on the ZnO concentration. Nano ZnO has no significant effect on the melting temperature, ΔH_m, and degree of PP crystallinity (Table S2, Figures S7 and S8). The literature dealing with the effect of ZnO on the PP crystallinity are dubious since some authors reported on increased PP crystallinity [26,32], while others observed no changes, or even the opposite effect [29]. Similar to the case of HDPE/ZnO composites, only minimal changes in the degree of PP matrix crystallinity and a rather high degree of ZnO aggregation (Figure 1) are the explanations for the small effect of added nano ZnO on the mechanical properties of PP nanocomposites

(Table S2). ZnO addition to the PP can cause various effects on its crystallinity, depending on the composite preparation process; consequently, different effects on the mechanical properties are reported. Some authors reported on enhancement of mechanical properties [22,26,27], while others reported on negligible effect of added nano ZnO [29,38].

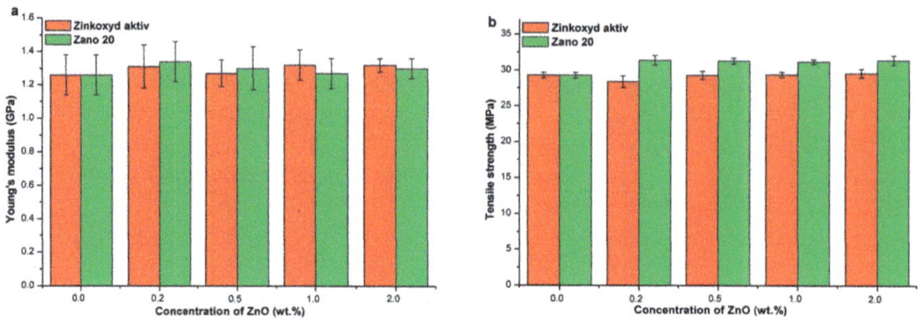

Figure 4. Mechanical properties of PP/ZnO composites as a function of nano ZnO concentration: (**a**) Young's modulus and (**b**) Tensile strength.

Many research groups also studied the effect of ZnO on the UV stability of polymer matrices. Literature reports enhanced polymer UV-stability since ZnO is an excellent UV absorber, however, the absorbed energy can be transferred to the polymer chains, causing their scission. Upon UV light exposure, ZnO is excited, leading to the formation of oxygen-active (hydroperoxide) species in the presence of OH groups on the surface, which can cause scission of polymer chains [17,39]. The first change caused by the UV light is the color or gloss change [40]. Therefore, the color change of polyolefin/ZnO nanocomposites was measured as a function of the exposure time of the composite to the artificial sun light.

Figures S9 and S10 show changes in the color (ΔE) of HDPE/ZnO composites (Zinkoxyd aktiv—Figure S9 and Zano 20—Figure S10), depending on the exposure time to the artificial sunlight. ΔE up to 7 weeks does not exceed the value of 4, meaning that with the naked eye, the color change is barely detectable. After ten weeks of exposure, ΔE reached the values from 8 to 15, meaning that color change was easily detected by naked eye. The change in color also depends on the nano ZnO concentration, since the largest color changes were observed at the highest nano ZnO concentration (2.0 wt%). Surface modification of nano ZnO with stearic acid reduces color changes (Figures S9 and S10). A comparison of both ZnO nanoparticles revealed that Zano 20 has a more pronounced effect on the color change of HDPE/ZnO composites than Zinkoxyd aktiv, which is in line with the presence of organic layer on the surface of Zinkoxyd aktiv. Other authors reported on similar effects of ZnO and TiO_2 particles on the UV stability of PE [19,39].

Figure S11 shows a color change of the PP/ZnO composites as a function of exposure time to the artificial sunlight. ΔE does not exceed 4, meaning only a moderate color change over ten weeks of exposure, which is equal to two years in real sunlight. The color change depends on the nano ZnO concentration, since the largest change was observed at 2.0 wt% of added ZnO. Surface modification of nano ZnO with stearic acid reduces the effect of nano ZnO, similar to the case of HDPEP/ZnO composites (Figure S11). Compared to HDPE, the PP matrix shows a higher stability to the artificial sunlight (Figure S9), indicating that a lifetime of PP/ZnO composites significantly exceeds two years even at 2.0 wt% of nano ZnO (Zinkoxyd aktiv).

HDPE/ZnO and PP/ZnO composites aged under artificial sunlight were studied by FTIR-ATR spectroscopy to perceive possible chemical processes and changes in chemical composition that occurred in the material (Figures S12 and S13). The FTIR spectra in Figure S12A–C show only minor differences compared to those of neat HDPE treated in the same way, leading to a conclusion that observed color changes (Figures S9 and S10) are not in correlation with the chemical changes (degradation) of the studied HDPE/ZnO composite samples. A comparison of FTIR spectra of PP/ZnO composites without and with 1.0 wt% of ZnO exposed to sunlight (Figure S13A–C) show the appearance of a new absorption band at 1725 cm^{-1}, corresponding to the formation of carbonyl groups. We studied its intensity as a function of composite exposure time (Figure S14). Results reveal increasing intensity of the carbonyl absorption band with time, but differences between the pure PP and PP/ZnO composites (unmodified and modified) are negligible, so no correlation can be established with the ZnO concentration in the PP matrix. We concluded that carbonyl absorption band is not related to the presence of ZnO [40].

2.3. *Composites of Nano ZnO with HDPE and PP Matrix—ZnO Surface Modified with Silanes*

HDPE and PP composites were prepared also with commercially available nano ZnO powders, the surface of which was modified with silanes (Zano 20 Plus: 3.9 wt% of caprylyl silane and Zano 20 Plus 3: 1.0 wt% of methacrylic silane). SEM micrographs in Figure S15 show distributions of silanized ZnO particles in composites based on HDPE and PP matrices. In comparison with composites prepared with unmodified ZnO and ZnO modified with stearic acid (Figures S1 and S5), the micrographs in Figure S15 show the presence of significantly smaller ZnO aggregates (size below 2 μm), confirming that silanization of ZnO considerably improves compatibility between the polyolefin matrix and nano ZnO.

Table S3 summarizes the mechanical properties of HDPE/silanized ZnO. When Zano 20 Plus nanofiller was used, a slight increase in tensile strength (by up to 4.7%—Figure 5a, Figure S16A) and elongation at break was observed, while Young's modulus is somewhat reduced. In the case of Zano 20 Plus 3, only elongation at break slightly increased, while tensile strength and Young's modulus were slightly reduced. The reduction of Young's modulus is attributed to the presence of silane, acting as the plasticizer. Overall, silanized ZnO has only a minor effect on the mechanical properties of ZnO composites with HDPE matrix as indicated by enhancement of Young's modulus and tensile strength by 2.0% and 4.7%, respectively. These results are in accordance with the composites' thermal properties, showing no relationship with ZnO concentration (Table S3). Some authors, on the other hand, reported on the enhancement of composites' mechanical properties when ZnO was modified with the silane coupling agents [7,31].

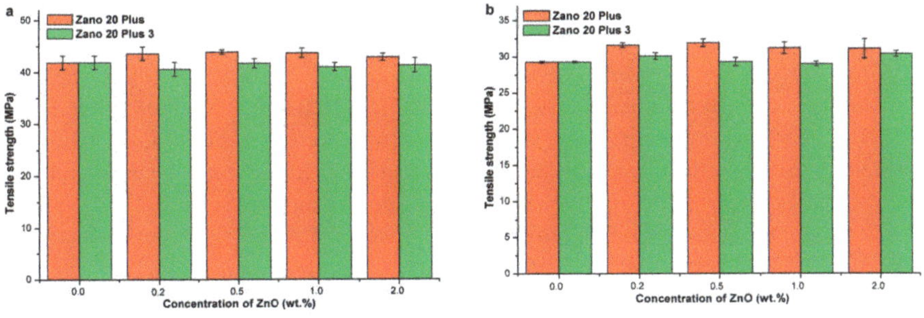

Figure 5. Tensile strength of polyolefin/ZnO composites as a function of silanized nano ZnO concentration (Zano 20 Plus): (**a**) HDPE and (**b**) PP matrix.

Table S4 summarizes the mechanical properties of PP/silanized ZnO composites. When Zano 20 Plus was used, the tensile strength (by up to 8.7%—Figure 5b, Figure S16B) and elongation at break increased, while Young's modulus slightly decreased. In the case of Zano 20 Plus 3, only elongation at break increases, while tensile strength and Young's modulus are more or less unaffected. Overall, the silanized ZnO has a rather small influence on the mechanical properties of composites based on the PP matrix since the maximal enhancement of tensile strength is 8.7%. Young's modulus is reduced in most cases, which is attributed to the plasticizing effect of silane. The thermal properties of PP/silanized ZnO composites showed a substantial increase (14.3%) in melting enthalpy when Zano 20 Plus was used (Table S4, Figure S17), while composites with Zano 20 Plus 3 showed no significant changes.

These observations are in accordance with an increase in tensile strength (Table S4) when Zano 20 Plus nanofiller was used, confirming that an increase in crystallinity degree is a predominant mechanism of mechanical properties reinforcement. These results are in contrast to those results, reporting on negligible changes of mechanical properties when silane coupling agents were applied [27,30], but in line with those reporting on enhancement of composite mechanical properties when nano ZnO surface modified with methacrylic silane was applied [32].

Figures S18 and S19 show color changes of HDPE/ZnO and PP/ZnO composites (Zano 20 Plus and Zano 20 Plus 3), depending on the composite exposure time to artificial sunlight and ZnO concentration. The change in color strongly depends on the ZnO concentration, since it is the greatest at 2.0 wt% (Figures S18 and S19). Moreover, a thicker layer of silane (Zano 20 Plus—3.9 wt% of caprylyl silane) slightly reduces color change as compared to the thinner layer (Zano 20 Plus 3–1.0 wt% methacrylic silane). A comparison of color changes shows that HDPE is significantly more light-sensitive than PP, since the latter shows only moderate changes in the color metrics after 10 weeks of exposure time; ΔE is below 4 for PP and between 6 and 11 for HDPE (Figures S18 and S19).

2.4. Antibacterial Properties of Polyolefin Composites with ZnO Nanoparticles

ZnO is well known as a highly efficient antibacterial agent. The exact mechanism of ZnO antibacterial activity is still not completely elucidated. There are many possible physical and chemical mechanisms of the ZnO interaction with the bacterial cells: (a) generation of reactive oxygenated species (ROS = $O_2^{-\bullet}$, $HO_2^{\bullet}$ and $HO^{\bullet}$) under UV or visible light illumination; (b) release of Zn^{2+} ions due to partial dissolution of ZnO; (c) generation of H_2O_2 by photoinduction; (d) disruption of plasma membrane due to interaction with ZnO; (e) internalization (penetration) of ZnO nanoparticles into the bacterial cell; (f) mechanical damage of the cell membrane [41]. The first three interaction mechanisms are of a chemical nature, while the other three are physical. The real effect of ZnO on the bacterial cells is most probably a combination of these mechanisms.

The measurements of antibacterial activity were carried out on the selected samples (Table 1) of ZnO composites with polyolefin matrices (Table 2, Table 3). We selected the samples containing 2.0 wt% of ZnO either Zinkoxyd aktiv or Zano 20, and ZnO of different surface modification. According to ISO 22196: 2007 standard, the log of reduction must be equal to 2 or higher than 2 so that certain material can be accepted as antibacterial [42]. The results show that all composites have excellent antibacterial activity against *S. aureus* (gram-positive bacteria) (Table 3), while the results on antibacterial activity against *E. coli* (gram-negative bacteria) vary significantly (Table 2) [41]. It is known that ZnO shows a more intense effect on the gram-positive bacteria such as *S. aureus* or *Bacillus subtilis* than on the gram-negative bacteria such as *E. coli* or *Aerobacter aerogenes* [25,43–45]. This can be attributed to their structural and compositional differences. Namely, gram-negative bacteria have an additional outer plasma membrane that consists of a thick lipopolysaccharide layer. The overall thickness of the membrane is larger in the gram-negative bacteria than in the gram-positive ones. These structural differences are the most probable reason for higher resistance of gram-negative bacteria towards ZnO [43–45]. Photographs of bacterial colonies on agar (Figure 6), taken after 24 h of contact with HDPE/ZnO composites containing 2.0 wt% of unmodified nano ZnO (Figure 6B,C), present a significant reduction in the number of *E. coli* bacterial colonies.

Table 1. Data on samples of HDPE or PP/ZnO composites selected for antibacterial activity testing (the concentration of nano ZnO is 2.0 wt%). Zinkox-aktiv = Zinkoxyd aktiv.

Designation of Samples	Polymer Matrix	Type of Nano ZnO	Type of Modification	Amount of Modifier (wt%)
PEZnO-1	HDPE	Zinkox-aktiv	-	-
PEZnO-2	HDPE	Zano 20	-	-
PEZnO-3	HDPE	Zano 20	Stearic acid	3.0
PEZnO-4	HDPE	Zano 20 Plus	Caprylyl silane	3.9
PPZnO-1	PP	Zinkox-aktiv	-	-
PPZnO-2	PP	Zano 20	-	-
PPZnO-3	PP	Zinkox-aktiv	Stearic acid	3.0

Table 2. Determined antibacterial activities and adequate activities scores of HDPE/ZnO and PP/ZnO composites against *E. coli*. Zinkox-aktiv = Zinkoxyd aktiv.

Designation of Samples	Type of Nano ZnO	Ref. (t = 0) [CFU/mL]	Ref. (t = 24 h) [CFU/mL]	Sample (t = 24 h) [CFU/mL]	U_0	U_t	A_t ($\log_{10}$ (CFU/mL))	R	Score
PEZnO-1	Zinkox-aktiv	3.9×10^3	3.6×10^3	3.1×10^1	3.39	3.35	1.29	2.06	Good
PEZnO-2	Zano 20	3.9×10^3	3.6×10^3	1	3.39	3.35	−0.20	3.55	Excellent
PEZnO-3	Zano 20	3.9×10^3	3.6×10^3	4.2×10^2	3.39	3.35	2.42	0.92	Poor
PEZnO-4	Zano 20 Plus	3.9×10^3	3.6×10^3	1.3×10^3	3.39	3.35	2.91	0.44	Bad
PPZnO-1	Zinkox-aktiv	3.8×10^3	3.6×10^3	4.4×10^2	3.38	3.35	2.44	0.91	Poor
PPZnO-2	Zano 20	3.8×10^3	3.6×10^3	3	3.38	3.35	0.27	3.08	Excellent
PPZnO-3	Zinkox-aktiv	3.8×10^3	3.6×10^3	5.9×10^2	3.38	3.35	2.56	0.78	Poor

Table 3. Determined antibacterial activities and adequate activities scores of HDPE/ZnO and PP/ZnO composites against *S. aureus*. Zinkox-aktiv = Zinkoxyd aktiv.

Designation of Samples	Type of Nano ZnO	Ref. (t = 0) [CFU/mL]	Ref. (t = 24 h) [CFU/mL]	Sample (t = 24 h) [CFU/mL]	U_0	U_t	A_t ($\log_{10}$(CFU/mL))	R	Score
PEZnO-1	Zinkox-aktiv	1.1×10^4	1.2×10^4	1.5	3.83	3.89	−0.03	3.91	Excellent
PEZnO-2	Zano 20	1.1×10^4	1.2×10^4	0.5	3.83	3.89	−0.50	4.39	Excellent
PEZnO-3	Zano 20	1.1×10^4	1.2×10^4	1	3.83	3.89	−0.03	4.09	Excellent
PEZnO-4	Zano 20 Plus	1.1×10^4	1.2×10^4	2	3.83	3.89	0.10	3.79	Excellent
PPZnO-1	Zinkox-aktiv	8.7×10^3	8.4×10^3	1	3.73	3.72	−0.20	3.92	Excellent
PPZnO-2	Zano 20	8.7×10^3	8.4×10^3	0.5	3.73	3.72	−0.50	4.22	Excellent
PPZnO-3	Zinkox-aktiv	8.7×10^3	8.4×10^3	1.5	3.73	3.72	−0.03	3.75	Excellent

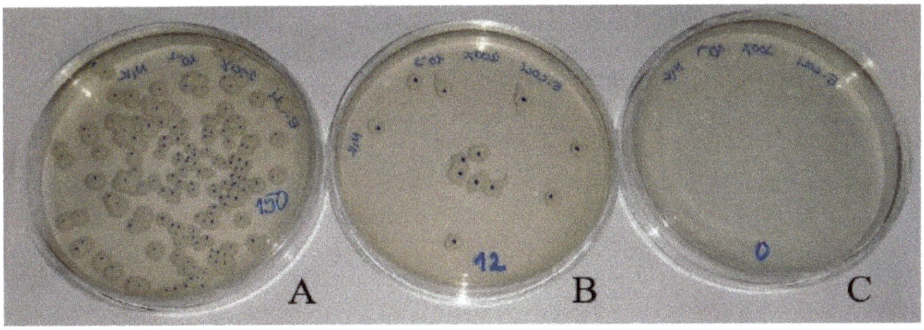

Figure 6. Determination of *E. coli* cell number (CFU/mL): (**A**) Starting inoculum concentration of *E. coli*; (**B**,**C**) *E. coli* count on two parallel test samples of HDPE/unmodified ZnO composite after 24 h.

Samples of surface unmodified ZnO show good or excellent activities, while those of surface modified ZnO are poor or even bad (Table 2). Comparing both ZnO nanofillers, Zano 20 shows higher antibacterial activity than Zinkoxyd aktiv, but the difference is mainly in the case of *E. coli*. Based on these results, we conclude that for achieving sufficient antibacterial activities towards both types of bacteria, the application of commercial unmodified Zano 20 ZnO in concentration of 2.0 wt% is recommended, although unmodified Zinkoxyd aktiv also show sufficient antibacterial activity with polyolefin matrices. The observed differences between *E. coli* and *S. aureus* in the antibacterial effect of polyolefin/ZnO composites originate from different membrane structure of gram-negative and gram-positive bacteria. Concerning the possible interaction mechanisms between ZnO and bacteria, we assume that chemical mechanisms do not require a physical contact between the ZnO particle and bacterial membrane since ROS, Zn^{2+} ions, and H_2O_2 can migrate from the surface of ZnO to the surface of bacteria. We conclude that the membrane structure of *S. aureus* allows the penetration of antibacterial active chemical compounds into the cell, thus explaining why, in this case, surface modified ZnO is also effective (Table 2). On the other side, the cell of *E. coli* is most probably damaged or destroyed only by the physical processes, requiring physical contact with ZnO, which thus explains why, in this case, only the unmodified ZnO is effective against this type of bacteria (Table 2).

3. Materials and Methods

3.1. Materials

Commercially available TIPELIN BA 550-13 HDPE and TIPPLEN K-499 PP granules, and ZnO nanopowders: Zinkoxyd aktiv, Lanxsess, Germany and Zano 20, Umicore, Belgium. In addition, commercial silane modified Nano ZnO nanofillers were also applied: Zano 20 Plus, Zano 20 Plus 3, Umicore, Belgija. The Zano 20 Plus was surface modified with 3.9 wt% of triethoxycaprylylsilane caprylylsilane, while the Zano 20 Plus 3 was modified with 1 wt% of [3-methacryloxypropyl]trimethoxysilane methacrylic silane. *Escherichia coli* strain (DSM 498) and *Staphilococcus aureus* strain (DSM 346) were supplied by DSMZ-Deutsche Sammlung von Mikroorganismen und Zellkulturen GmbH, Germany.

3.2. Preparation of Polyolefin/ZnO Composites

ZnO nanopowders were applied to the surface of the PE or PP granulate. For this purpose, nano ZnO was first suspended in ethanol and subsequently the PE or PP granulate was added and sonicated. Finally, ethanol was evaporated on the rotary evaporator. The prepared granulates were extruded at 160 °C (PE) or 180 °C (PP) for 10 min at 50 rpm with Haake MiniLab extruder (Thermo Fischer Scientific, Karlsruhe, Germany). The mixture was added to the extruder in two portions of 3 g. The extruded melt was ejected from the extruder at 100 rpm and it was captured in a heated container (170 °C), which was further placed into a Haake Mini Jet molding machine (Thermo Fischer Scientific, Karlsruhe, Germany) to prepare the testing specimens by injection into a suitable mold heated to 70 °C at the pressure of 750 bar and time of 10 s, as well as the post pressure of 250 bar and time of 10 s.

3.3. Characterization

Mechanical properties were measured according to ISO 527 standard on the Shimadzu AGS-GX (Shimadzu, Kyoto, Japan) plus a dynamometer with an initial spacing of 58 mm and the stretching speed of 2 mm/min and 200 mm/min.

UV-vis resistance of composites was determined by exposing the samples to artificial sunlight in the Suntest chamber for certain time periods. Color change was measured as a function of time by using an i1 Pro (X-Rite, Grand Rapids, MI, USA) spectrometer, measuring the spectra of reflectivity in a spectral range of 380–730 nm.

Fourier-transform infrared spectroscopy (FTIR) spectra were recorded in a transmittance mode on a FTIR spectrometer Spectrum One (Perkin-Elmer, Waltham, MA, USA) in a spectral range between 360 cm^{-1} and 4000 cm^{-1}, and a spectral resolution of 4 cm^{-1} using KBr tablets (ZnO powders) or in an ATR mode in a spectral range from 650 to 4000 cm^{-1} with a 4 cm^{-1} spectral resolution (polyolefin/ZnO composites).

Photoluminescence spectra of ZnO powders were recorded on a Perkin Elmer LS-55 spectrometer (Perkin-Elmer, Waltham, MA, USA) in a range from 330 nm to 620 nm using an excitation wavelength of 325 nm.

SEM micrographs of composites were taken on a Zeiss Supra 35 VP field emission electron microscope (Zeiss, Oberkochen, Germany) at a 15 kV acceleration voltage using a back-scattered electron detector at a working distance of 8 mm.

Thermal properties of polyolefin matrices and composites were determined by DSC calorimetry on DSC-1 (Mettler Toledo, Greifensee, Switzerland) in a temperature range from 25 °C to 200 °C at a heating rate of 10 K/min.

Thermal stability of composite samples was determined by thermogravimetric analysis (TGA) in oxygen atmosphere using a TGA-1 (Mettler Toledo, Greifensee, Switzerland) instrument. The measurements were performed in a temperature range from 30 °C to 800 °C at a heating rate of 10 K/min.

Antibacterial activity of ZnO/polyolefin composites was determined according to ISO 22196: 2007 standard.

Crystalline fractions of the ZnO powders were characterized by a wide-angle X-ray diffraction (XRD) on an XPert Pro diffractometer (PANalytical, Almelo, Netherlands) with Cu anode as an X-ray source. X-ray diffractograms were measured at 25 °C in the 2θ range from 5° to 80° with a step of 0.033° and step time of 100 s. Crystallite sizes were calculated using the Scherrer formula [10] and Si wafer was used to determine the experimental peak broadening.

Nitrogen sorption measurements were performed on a manometric gas sorption analyzer (Micromeritics Instrument Co., Norcross, GA, USA) at −196 °C in the range of relative pressure values from 10^{-6} to 1. As-prepared samples were degassed at 140 °C for 16 h prior to the measurements. The specific surface areas were determined by BET method based on the obtained sorption isotherms.

4. Conclusions

SEM microscopy revealed agglomerated nano ZnO with sizes from 1 to 5 μm in polyolefin matrices. Results of tensile testing showed that both nano ZnO do not enhance the mechanical properties of HDPE composites, while PP composites show slight enhancement (Young's module by 6% and tensile strength by 7%) when Zano 20 was used as a nanofiller. Measurements of thermal properties revealed only a small effect of nano ZnO on the degree of crystallinity of these composites, which is in accordance with the unchanged mechanical properties. Surface modification of ZnO with stearic acid increases its compatibility with polyolefin matrices, which, however, does not result in improved composite mechanical properties. Color change of composite materials was studied in dependence of ZnO concentration, exposure time to artificial sunlight, and ZnO type. Higher nano ZnO concentrations cause more obvious color changes of HDPE/ZnO composites, while surface modification of ZnO with stearic acid partially reduces this effect. Nevertheless, the PP matrix shows significantly smaller changes (barely visible with naked eye—ΔE is below 4) than the HDPE matrix (ΔE is between 8 and 15) after 10 weeks of exposure of composites to artificial sunlight. As revealed by TGA, added nano ZnO (both ZnO types—2.0 wt%) also increased the thermal stability of HDPE while no changes were observed for PP matrix.

Silanization of nano ZnO significantly improved compatibility of ZnO with polyolefin matrices as indicated by smaller ZnO agglomerates (1–2 μm). Despite better ZnO distribution in polyolefin matrices, it has only a minor effect on the composites' mechanical properties. Nevertheless, clear positive trends were observed with silanized nano ZnO as indicated by 5–9% enhanced tensile strength when caprylyl silane (3.9 wt%) was applied as the ZnO surface modifying agent in the PP matrix. This was accompanied by 14.3% increase in the PP melting enthalpy, confirming that increased PP crystallinity is responsible for the enhanced composite mechanical properties. Young's modulus is reduced, which is attributed to the plasticizing effect of silane. Silanization also slightly reduces color changes of polyolefin/ZnO composites. Concerning improvement of mechanical properties in general, the use of nano ZnO powders as the reinforcing agents in polyolefins is economically not justified.

The composites with unmodified and surface modified ZnO are highly active against *S. aureus*, while in the case of *E. coli*, only the composites with unmodified ZnO show sufficient antibacterial activity. This difference is attributed to differences in membrane structure between the gram-positive (*S. aureus*) and gram-negative (*E. coli*) bacteria. To achieve sufficient antibacterial activity of polyolefin composites towards both bacteria types, the application of unmodified commercial ZnO (Zano 20) in the concentration of 2.0 wt% is recommended. Incorporation of ZnO into the polyolefin matrices is considered to be a promising way to improve material antibacterial properties for clinical application or applications in the food industry.

Supplementary Materials: The following Figures and Tables are available online, Figure S1: SEM micrographs (magnification 1000×, backscattered electron detector) of HDPE/ZnO composites (Zinkoxyd aktiv – 1.0 wt%): (A) unmodified nano ZnO - Zinkoxyd aktiv; (B) nano ZnO - Zinkoxyd aktiv modified with stearic acid; (C) unmodified nano ZnO - Zano 20; (D) nano ZnO-Zano 20 modified with stearic acid, Table S1: Mechanical and thermal properties of HDPE/ZnO (Zinkoxyd aktiv or Zano 20) composites as the functions of nano ZnO concentration and surface modification of ZnO with stearic acid. The polymer matrix is polyethylene HDPE - H-3581, the addition of stearic acid is 3% by weight, Znoxaktiv is Zinkoxyd-aktiv, Figure S2: Stress-strain curves of HDPE/ZnO composites as a function of ZnO concentration: 0 wt% - black; 0.5 wt% - green; 2.0 wt% blue: (A) Zinkoxyd aktiv; (B) Zano 20, Figure S3: DSC curves of HDPE/ZnO composites as a function of ZnO concentration, Figure S4: TGA curves of HDPE/ZnO composites as a function of ZnO concentration, Figure S5: SEM micrographs (magnification 1000×, backscattered electron detector) of PP/ZnO composites (Zinkoxyd aktiv – 1.0 wt%): (A) unmodified nano ZnO - Zinkoxyd aktiv; (B) nano ZnO - Zinkoxyd aktiv modified with stearic acid; (C) unmodified nano ZnO - Zano 20; (D) nano ZnO-Zano 20 modified with stearic acid, Table S2: Mechanical and thermal properties of composites with PP matrix and ZnO - Zinkoxyd aktiv or Zano 20 (unmodified ZnO and surface modified with stearic acid) depending on the nano ZnO concentration. The polymer matrix is polypropylene PP - K499, the addition of stearic acid is 3.0 wt%, Znox-aktiv is Zinkoxyd aktiv, Figure S6: Stress-strain curves of PP/ZnO composites as a function of ZnO concentration: 0 wt% - black; 0.5 wt% - green; 2.0 wt% blue: (A) Zinkoxyd aktiv; (B) Zano 20, Figure S7: DSC curves of PP/ZnO composites as a function of ZnO concentration, Figure S8: TGA curves of PP/ZnO composites as a function of ZnO concentration, Figure S9: Changes in color of HDPE/ZnO composites (Zinkoxyd aktiv) as the functions of exposure time to artificial sunlight, nano ZnO concentration, and ZnO surface modification with stearic acid, Figure S10: Changes in color of HDPE/ZnO composites (Zano 20) as the functions of exposure time to artificial sunlight, nano ZnO concentration, and ZnO surface modification with stearic acid, Figure S11: Changes in color of PP/ZnO composites (Zinkoxyd aktiv) as the functions of exposure time to the artificial sunlight and ZnO concentration, Figure S12: FTIR spectra of HDPE/ZnO composites prior to exposure to artificial light and after ten weeks of exposure: (A) pure HDPE; (B) HDPE with 1.0 wt% of ZnO (Zinkoxyd-aktiv); (C) HDPE with 1.0 wt% of ZnO (Zano 20), Figure S13: FTIR spectra of PP/ZnO composites prior to exposure to artificial light and after ten weeks of exposure: (A) pure PP; (B) PP with 1.0 wt% of ZnO (Zinkoxyd-aktiv); (C) PP with 1.0 wt% of ZnO (Zano 20), Figure S14: Intensity of the absorption band at 1725 cm^{-1} in PP/ZnO composites as a function of exposure time to the artificial sunlight, Figure S15: SEM micrographs (magnification 1000×, backscattered electron detector) of HDPE/ZnO composites with 1.0 wt% of silane modified ZnO: (A) Zano 20 Plus; (B) Zano 20 Plus 3, and PP/ZnO composites: (C) Zano 20 Plus; (D) Zano 20 Plus 3, Table S3: Mechanical and thermal properties of composites of HDPE matrix and silanized nano ZnO (Zano 20 Plus - 3.9 wt% of caprylyl silane and Zano 20 Plus 3 – 1.0 wt% of methacrylic silane), Table S4: Mechanical and thermal properties of composites with PP matrix and silanized nano ZnO (Zano 20 Plus – 3.9 wt% of caprylyl silane and Zano 20 Plus 3 – 1.0 wt% of methacrylic silane), Figure S16: Stress-strain curves of polyolefin/ZnO composites as a function of silanized ZnO concentration (Zano 20 Plus): 0 wt% - black; 0.5 wt% - green; 2.0 wt% blue. (A) HDPE and (B) PP, Figure S17: DSC curves of PP/ZnO composites as a function of silanized ZnO concentration (Zano 20 Plus), Figure S18: Change in color of HDPE/ZnO composites (Zano 20 Plus and Zano 20 Plus 3) depending on the exposure time to artificial sunlight and nano ZnO concentration as well as the type of surface modification, Figure S19: Change in color of PP/ZnO composites (Zano 20 Plus and Zano 20 Plus 3) as the functions of the exposure time to artificial sunlight and nano ZnO concentration as well as the type of surface modification.

Author Contributions: Conceptualization, A.A. and I.Š.; Methodology, A.A., I.Š., and M.P.; Validation, A.A., M.P., I.Š., M.L., Ž.K., and E.Ž.; Formal Analysis, A.A., I.Š., M.P.; Investigation, A.A. and M.P.; Data Curation, M.L. and E.Ž.; Writing—Original Draft Preparation, A.A.; Writing—Review and Editing, A.A., M.P., I.Š., M.L., Ž.K., and E.Ž.; Visualization, A.A.; Supervision, M.L., Ž.K., and E.Ž.; Project Administration, M.L., Ž.K., and E.Ž.; Funding Acquisition, Ž.K. and E.Ž.

Funding: This research was funded by the SLOVENIAN RESEARCH AGENCY (Research Core Funding No. P2-0145 and P2-0046) and the SLOVENIAN MINISTRY OF EDUCATION, SCIENCE, AND SPORT (Program Martina No. OP20.00369).

Acknowledgments: Authors acknowledge the support of Marta Klanjšek Gunde for the access to equipment for color change measurements.

Conflicts of Interest: The authors declare no conflict of interest.

References

1. Jancar, J.; Douglas, J.F.; Starr, F.W.; Kumar, S.K.; Cassagnau, P.; Lesser, A.J.; Sternstein, S.S.; Buehler, M.J. Current issues in research on structure–property relationships in polymer nanocomposites. *Polymer* **2010**, *51*, 3321–3343. [CrossRef]
2. De Salzano, M.S.; Galizia, M.; Wojnarowicz, J.; Rosa, R.; Lojkowski, W.; Leonelli, C.; Acierno, D.; Filippone, G. Dispersing hydrophilic nanoparticles in hydrophobic polymers: HDPE/ZnO nanocomposites by a novel template-based approach. *eXPRESS Polym. Lett.* **2014**, *8*, 362–372. [CrossRef]
3. Balazs, A.C.; Emrick, T.; Russell, T.P. Nanoparticle Polymer Composites: Where Two Small Worlds Meet. *Science* **2006**, *314*, 1107–1110. [CrossRef] [PubMed]
4. Demir, M.M.; Wegner, G. Challenges in the Preparation of Optical Polymer Composites With Nanosized Pigment Particles: A Review on Recent Efforts. *Macromol. Mater. Eng.* **2012**, *297*, 838–863. [CrossRef]

5. Peponi, L. Morphology-properties relationship on nanocomposite films based on poly(styrene-block-diene-block-styrene) copolymers and silver nanoparticles. *Express Polym. Lett.* **2011**, *5*, 104–118. [CrossRef]
6. Chow, W.S. Effects of SEBS-g-MAH on the properties of injection moulded poly(lactic acid)/nano-calcium carbonate composites. *Express Polym. Lett.* **2012**, *6*, 503–510. [CrossRef]
7. Li, Y.-N.; Li, S.-C.; Li, S. Mechanical and antibacterial properties of modified nano-ZnO/high-density polyethylene composite films with a low doped content of nano-ZnO. *J. Appl. Polym. Sci.* **2010**, *116*, 2965–2969. [CrossRef]
8. Pahovnik, D.; Orel, Z.C.; Kos, T.; Anžlovar, A.; Žagar, E.; Žigon, M. Zinc-Containing Block Copolymer as a Precursor for the in situ Formation of Nano ZnO and PMMA/ZnO Nanocomposites. *Macromolecules* **2013**, *46*, 6942–6948.
9. Anzlovar, A. Polyol mediated nano size zinc oxide and nanocomposites with poly(methyl methacrylate). *Express Polym. Lett.* **2011**, *5*, 604–619. [CrossRef]
10. Anžlovar, A.; Marinšek, M.; Orel, Z.C.; Žigon, M. Basic zinc carbonate as a precursor in the solvothermal synthesis of nano-zinc oxide. *Mater. Des.* **2015**, *86*, 347–353. [CrossRef]
11. Yang, Z.; Zong, X.; Ye, Z.; Zhao, B.; Wang, Q.; Wang, P. The application of complex multiple forklike ZnO nanostructures to rapid and ultrahigh sensitive hydrogen peroxide biosensors. *Biomaterials* **2010**, *31*, 7534–7541. [CrossRef] [PubMed]
12. Kogej, K.; Orel, Z.C.; Anžlovar, A.; Žigon, M. Impact of Inorganic Hydroxides on ZnO Nanoparticle Formation and Morphology. *Cryst. Growth Des.* **2014**, *14*, 4262–4269.
13. Anžlovar, A.; Orel, Z.C.; Kogej, K.; Žigon, M. Polyol-Mediated Synthesis of Zinc Oxide Nanorods and Nanocomposites with Poly(methyl methacrylate). *J. Nanomater.* **2012**, *2012*, 1–9. [CrossRef]
14. Djurišić, A.B.; Leung, Y.H. Optical Properties of ZnO Nanostructures. *Small* **2006**, *2*, 944–961. [CrossRef] [PubMed]
15. Abou-Kandil, A.I.; Awad, A.; Mwafy, E. Polymer nanocomposites part 2: Optimization of zinc oxide/high-density polyethylene nanocomposite for ultraviolet radiation shielding. *J. Thermoplast. Comp. Mater.* **2015**, *28*, 1583–1598. [CrossRef]
16. Yang, R.; Li, Y.; Yu, J. Photo-stabilization of linear low density polyethylene by inorganic nano-particles. *Polym. Degrad. Stab.* **2005**, *88*, 168–174. [CrossRef]
17. Podbršček, P.; Dražič, G.; Anžlovar, A.; Orel, Z.C. The preparation of zinc silicate/ZnO particles and their use as an efficient UV absorber. *Mater. Res. Bull.* **2011**, *46*, 2105–2111. [CrossRef]
18. Anžlovar, A.; Kržan, A.; Žagar, E. Degradation of PLA/ZnO and PHBV/ZnO composites prepared by melt processing. *Arab. J. Chem.* **2018**, *11*, 343–352. [CrossRef]
19. Yang, R.; Christensen, P.; Egerton, T.; White, J. Degradation products formed during UV exposure of polyethylene–ZnO nano-composites. *Polym. Degrad. Stab.* **2010**, *95*, 1533–1541. [CrossRef]
20. Sawai, J.; Igarashi, H.; Hashimoto, A.; Kokugan, T.; Shimizu, M. Effect of Ceramic Powder Slurry on Spores of Bacillus subtilis. *J. Chem. Eng. Jpn.* **1995**, *28*, 556–561. [CrossRef]
21. Sawai, J.; Yoshikawa, T. Quantitative evaluation of antifungal activity of metallic oxide powders (MgO, CaO and ZnO) by an indirect conductimetric assay. *J. Appl. Microbiol.* **2004**, *96*, 803–809. [CrossRef] [PubMed]
22. Ozmıhci, F.O.; Balkose, D. Effects of Particle Size and Electrical Resistivity of Filler on Mechanical, Electrical, and Thermal Properties of Linear Low Density Polyethylene–Zinc Oxide Composites. *J. Appl. Polym. Sci.* **2013**, *130*, 2734–2743. [CrossRef]
23. Hong, J.I.; Schadler, L.S.; Siegel, R.W.; Mårtensson, E. Rescaled electrical properties of ZnO/low density polyethylene nanocomposites. *Appl. Phys. Lett.* **2003**, *82*, 1956–1958. [CrossRef]
24. Wang, X.; Zhang, P.; Hong, R. Preparation and application of aluminum-doped zinc oxide powders via precipitation and plasma processing method. *J. Appl. Polym. Sci.* **2015**, *132*, 41990. [CrossRef]
25. Esmailzadeh, H.; Sangpour, P.; Shahraz, F.; Hejazi, J.; Khaksar, R. Effect of nanocomposite packaging containing ZnO on growth of Bacillus subtilis and Enterobacter aerogenes. *Mater. Sci. Eng. C* **2016**, *58*, 1058–1063. [CrossRef] [PubMed]
26. Esthappan, S.K.; Nair, A.B.; Joseph, R. Effect of crystallite size of zinc oxide on the mechanical, thermal and flow properties of polypropylene/zinc oxide nanocomposites. *Compos. Part B Eng.* **2015**, *69*, 145–153. [CrossRef]
27. Zhou, J.P.; Qiu, K.Q.; Fu, W.L. The Surface Modification of ZnOw and its Effect on the Mechanical Properties of Filled Polypropylene Composites. *J. Compos. Mater.* **2005**, *39*, 1931–1941. [CrossRef]

28. Helal, E.; Pottier, C.; David, E.; Frechette, M.; Demarquette, N. Polyethylene/thermoplastic elastomer/Zinc Oxide nanocomposites for high voltage insulation applications: Dielectric, mechanical and rheological behavior. *Eur. Polym. J.* **2018**, *100*, 258–269. [CrossRef]
29. Lepot, N.; Van Bael, M.K.; Van den Rul, H.; D'Haen, J.; Peeters, R.; Franco, D.; Mullens, J. Influence of Incorporation of ZnO Nanoparticles and Biaxial Orientation on Mechanical and Oxygen Barrier Properties of Polypropylene Films for Food Packaging Applications. *J. Appl. Polym. Sci.* **2011**, *120*, 1616–1623. [CrossRef]
30. Altan, M.; Yildirim, H. Effects of compatibilizers on mechanical and antibacterial properties of injection molded nano-ZnO filled polypropylene. *J. Compos. Mater.* **2012**, *46*, 3189–3199. [CrossRef]
31. Wang, Y.; Shi, J.; Han, L.; Xiang, F. Crystallization and mechanical properties of T-ZnOw/HDPE composites. *Mater. Sci. Eng. A* **2009**, *501*, 220–228. [CrossRef]
32. Zeng, A.; Zheng, Y.; Guo, Y.; Qiu, S.; Cheng, L. Effect of tetra-needle-shaped zinc oxide whisker (T-ZnOw) on mechanical properties and crystallization behavior of isotactic polypropylene. *Mater. Des.* **2012**, *34*, 691–698. [CrossRef]
33. Hayashi, S.; Nakamori, N.; Kanamori, H. Generalized Theory of Average Dielectric Constant and Its Application to Infrared Absorption by ZnO Small Particles. *J. Phys. Soc. Jpn.* **1979**, *46*, 176–183. [CrossRef]
34. Kleinwechter, H.; Janzen, C.; Knipping, J.; Wiggers, H.; Roth, P. Formation and properties of ZnO nano-particles from gas phase synthesis processes. *J. Mater. Sci.* **2002**, *37*, 4349–4360. [CrossRef]
35. Mishra, S.; Srivastava, R.; Prakash, S.; Yadav, R.S.; Panday, A. Photoluminescence and photoconductive characteristics of hydrothermally synthesized ZnO nanoparticles. *Opto-Electron. Rev.* **2010**, *18*, 467–473. [CrossRef]
36. Shalish, I.; Temkin, H.; Narayanamurti, V. Size-dependent surface luminescence in ZnO nanowires. *Phys. Rev. B* **2004**, *69*. [CrossRef]
37. Heller, R.B.; McGannon, J.; Weber, A.H. Precision Determination of the Lattice Constants of Zinc Oxide. *J. Appl. Phys.* **1950**, *21*, 1283. [CrossRef]
38. Silvestre, C.; Cimmino, S.; Pezzuto, M.; Marra, A.; Ambrogi, V.; Dexpert-Ghys, J.; Verelst, M.; Augier, S.; Romano, I.; Duraccio, D. Preparation and characterization of isotactic polypropylene/zinc oxide microcomposites with antibacterial activity. *Polym. J.* **2013**, *45*, 938–945. [CrossRef]
39. Allen, N.S.; Edge, M.; Corrales, T.; Catalina, F. Stabiliser interactions in the thermal and photooxidation of titanium dioxide pigmented polypropylene films. *Polym. Degrad. Stab.* **1998**, *61*, 139–149. [CrossRef]
40. Rouillon, C.; Bussiere, P.-O.; Desnoux, E.; Collin, S.; Vial, C.; Therias, S.; Gardette, J.-L. Is carbonyl index a quantitative probe to monitor polypropylene photo degradation? *Polym. Degrad. Stab.* **2016**, *128*, 200–208. [CrossRef]
41. Kumar, R.; Umar, A.; Kumar, G.; Nalwa, H.S. Antimicrobial properties of ZnO nanomaterials: A review. *Ceram. Int.* **2017**, *43*, 3940–3961. [CrossRef]
42. Altan, M.; Yildirim, H. Comparison of Antibacterial Properties of Nano TiO_2 and ZnO Particle Filled Polymers. *Acta Phys. Pol. A* **2014**, *125*, 645–647. [CrossRef]
43. Jain, A.; Bhargava, R.; Poddar, P. Probing interaction of Gram-positive and Gram-negative bacterial cells with ZnO nanorods. *Mater. Sci. Eng. C* **2013**, *33*, 1247–1253. [CrossRef] [PubMed]
44. Aal, N.A.; Al-Hazmi, F.; Al-Ghamdi, A.A.; Al-Ghamdi, A.; El-Tantawy, F.; Yakuphanoglu, F. Novel rapid synthesis of zinc oxide nanotubes via hydrothermal technique and antibacterial properties. *Spectrochim. Acta Part A Mol. Biomol. Spectrosc.* **2015**, *135*, 871–877. [CrossRef] [PubMed]
45. Tam, K.; Djurišić, A.; Chan, C.; Xi, Y.; Tse, C.; Leung, Y.; Chan, W.; Leung, F.; Au, D. Antibacterial activity of ZnO nanorods prepared by a hydrothermal method. *Thin Solid Films* **2008**, *516*, 6167–6174. [CrossRef]

Sample Availability: Samples of the compounds are not available from the authors.

molecules

MDPI

Article

Magnetic Polyurea Nano-Capsules Synthesized via Interfacial Polymerization in Inverse Nano-Emulsion

Suzana Natour [1], Anat Levi-Zada [2] and Raed Abu-Reziq [1,*]

[1] Institute of Chemistry, Casali Centre of Applied Chemistry and Centre for Nanoscience and Nanotechnology, The Hebrew University of Jerusalem, Jerusalem 9190401, Israel

[2] Department of Entomology-Chemistry, Agricultural Research Organization, Volcani Centre, Rishon Lezion 7505101, Israel

* Correspondence: Raed.Abu-Reziq@mail.huji.ac.il; Tel.: +972-2-6586097

Academic Editors: Marinella Striccoli, Roberto Comparelli and Annamaria Panniello

Received: 17 June 2019; Accepted: 18 July 2019; Published: 23 July 2019

check for updates

Abstract: Polyurea (PU) nano-capsules have received voluminous interest in various fields due to their biocompatibility, high mechanical properties, and surface functionality. By incorporating magnetic nanoparticle (MNPs) into the polyurea system, the attributes of both PU and MNPs can be combined. In this work, we describe a facile and quick method for preparing magnetic polyurea nano-capsules. Encapsulation of ionic liquid-modified magnetite nanoparticles (MNPs), with polyurea nano-capsules (PU NCs) having an average size of 5–20 nm was carried out through interfacial polycondensation between amine and isocyanate monomers in inverse nano-emulsion (water-in-oil). The desired magnetic PU NCs were obtained utilizing toluene and triple-distilled water as continuous and dispersed phases respectively, polymeric non-ionic surfactant cetyl polyethyleneglycol/polypropyleneglycol-10/1 dimethicone (ABIL EM 90), diethylenetriamine, ethylenediamine diphenylmethane-4,4′-diisocyanate, and various percentages of the ionic liquid-modified MNPs. High loading of the ionic liquid-modified MNPs up to 11 wt% with respect to the dispersed aqueous phase was encapsulated. The magnetic PU NCs were probed using various analytical instruments including electron microscopy, infrared spectroscopy, X-ray diffraction, and nuclear magnetic spectroscopy. This unequivocally manifested the successful synthesis of core-shell polyurea nano-capsules even without utilizing osmotic pressure agents, and confirmed the presence of high loading of MNPs in the core.

Keywords: polyurea nano-capsules; magnetic nanoparticles; nano-emulsions; interfacial polymerization; composite nanomaterials

1. Introduction

Nano-capsules, composed of liquid or hollow cores, enclosed in a nontoxic polymeric shell, have been widely investigated for the encapsulation of hydrophobic and hydrophilic substances [1]. Different methods for fabricating nano-capsules and nanoparticles have been developed. Typically, they can be synthesized through interfacial polymerization, suspension polymerization, and nanoprecipitation in oil-in-water (O/W), oil-in-oil (O/O), and water-in-oil (W/O) emulsions, as well as mini- and micro-emulsions [2–8]. Interfacial polymerization and polycondensation is one of the most studied methods for fabricating a wide range of functional polymeric nano-capsules. In this process, polymerization occurs at the interface between two immiscible phases, with each phase containing dissolved complementary monomers, resulting in nano-capsules and nanoparticles with sizes on the order of emulsion droplets [9,10].

To encapsulate hydrophilic compounds, an inverse nano-emulsion (W/O system) was utilized [11,12]. Inverse nano-emulsion, also known as inverse miniemulsion, consists of 50–500 nm surfactant-stabilized aqueous droplets dispersed in a hydrophobic organic continuous phase. Nano-emulsions are kinetically stable and their preparation requires high energy, for example, using high shear homogenization and ultra-sonication methods [13,14].

In order to obtain the desired polymeric nano-capsules, stable inverse nano-emulsions must be prepared. This is achieved by using a combination of non-ionic surfactants with low hydrophilic-lipophilic balance (HLB) and osmotic pressure agents (lipophobes) as co-stabilizers. The surfactant sterically stabilizes the droplets and lipophobes within the droplets, and prevents droplet coalescence by suppressing Ostwald ripening [15–17].

Among the nanostructured materials prepared via interfacial polymerization, polyurea has gained the most interest due to having properties such as biocompatibility, high mechanical characteristics, and surface functionality [14,18,19]. Only a few studies have reported the preparation of polyurea nano-capsules (PU NCs) by means of inverse miniemulsion by utilizing lipophobes. In this regard, Landfester and co-workers reported the preparation of hollow polyurea, polythiourea, as well as polyurethane nano-capsules and nanoparticles [20]. They studied the effect of monomers and solvents on the shell thickness and morphology to develop nanoreactors for preparing silver nanoparticles (NPs). In another study, PU NCs were stabilized with an amino-functionalized surfactant and the encapsulation efficiency was examined by a fluorescent dye [19].

Nano-capsules have a high surface area-to-volume ratio, a narrow size distribution, and high encapsulation efficiency. However, the isolation and recovery of such systems is difficult and requires time consuming and tedious procedures. Nevertheless, these efforts can be minimized by encapsulating magnetite nanoparticles (MNPs), which will endow the PU NCs with superparamagnetic properties. This process facilitates their isolation simply by applying an external magnetic field [21–24]. The polymeric shell protects the MNPs from undergoing agglomeration and oxidation, which otherwise leads to a loss of magnetic properties.

Previously, we reported the synthesis of magnetically separable PU nanoparticles formulated using O/O nano-emulsion by employing the interfacial polycondensation reaction between 2,6-diaminopyridine and polymethylene-polyphenyl isocyanate (PAPI 27) in the presence of poly(1-ethenylpyrrolidin-2-one/hexadec-1-ene) (Agrimer AL 22) surfactant [25]. Spherical particles of ~450 nm size were obtained. In line with that research, we proposed that magnetically separable PU NCs be prepared from inverse nano-emulsion. To the best of our knowledge, encapsulation of MNPs within PU NCs (MNPs@PU NCs), prepared from W/O nano-emulsions, has not been thoroughly investigated [19]. We believe that the encapsulated MNPs within polymeric capsules and matrixes, in which the attributes of PU and MNPs are combined, may be of great interest and can be applied in various fields such as biotechnology, medicine, catalysis, magnetic resonance imaging, agriculture, and other environmental and industrial applications [26–29]. Therefore, the aim of this work was to synthesize and provide an elaborative study as well as to thoroughly characterize new MNPs@PU NCs prepared in a facile manner from inverse nano-emulsion. In addition, we focused on studying the effect of different parameters, such as osmotic pressure agents, amine and isocyanate monomers, solvents, and surfactants.

Here, the magnetic polyurea nano-capsules were prepared through interfacial polycondensation in W/O nano-emulsion. The synthesis involved nano-emulsification of an aqueous phase containing ionic liquid (IL) stabilized magnetite nanoparticles, amine monomers, and an oil phase containing a polymeric non-ionic surfactant. This was followed by the addition of diisocyanate monomer to initiate the interfacial polycondensation, forming a polyurea shell and MNPs encapsulated within the core.

2. Results and Discussions

2.1. The Formation of Polyurea Nano-Capsules from Water-in-Oil (W/O) Nano-Emulsion

Prior to the encapsulation of the MNPs, an optimal composition for the synthesis of PU NCs was established through interfacial polymerization reactions in inverse nano-emulsions, as illustrated in Figure 1.

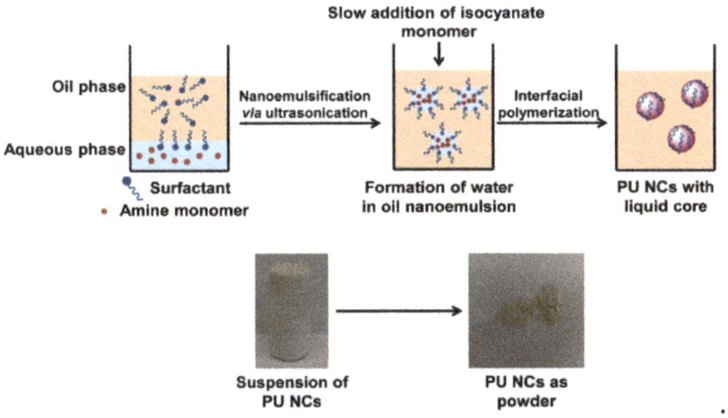

Figure 1. Preparation of polyurea nano-capsules from water-in-oil (W/O) nano-emulsion.

The polyurea nano-capsules were produced in two steps. Briefly, in the first step, the aqueous phase, composed of triple-distilled water (TDW), amine monomer, and lipophobe, was nano-emulsified by homogenization, followed by ultrasonication in an oil phase consisting of a non-polar organic solvent and surfactant. Subsequently, diisocyanate monomer was slowly added to the nano-emulsion system while sonication and the interfacial polycondensation were initiated to form the PU shell. The reaction between the amine and isocyanate monomers is depicted in Scheme 1.

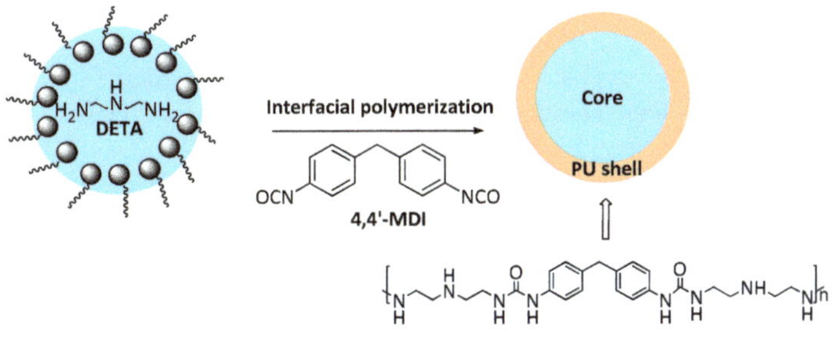

Scheme 1. The formation of polyurea through interfacial polymerization between diethylene triamine (DETA) and diphenylmethane-4,4′-diisocyanat (4,4′-MDI).

In order to attain stable water droplets dispersed in the oil phase, it is necessary to use the right surfactant. Thus, polymeric non-ionic surfactants with low HLB values were found to be most suitable, since they sterically stabilize the nanodroplets to prevent coalescence and provide a relatively condensed interface. Furthermore, an osmotic pressure agent, termed a lipophobe, was mainly used to maintain the stability of the nano-emulsion during the polymerization process [20]. This, hinders Ostwald ripening caused by nano-emulsion polydispersity and consequently, osmotic pressure forms inside the aqueous droplets, which reduces the Laplace pressure. Hence, it prevents the formation of aggregates.

In the above process, in order to obtain the optimal composition, various parameters in different proportions were tested, such as the type of surfactant, the amine and isocyanate monomers, the organic solvent, and the lipophobe in different ratios was also examined.

2.1.1. Variation of the Type and Amount of Surfactant

In preliminary experiments, different percentages of various surfactants with low HLB values were investigated using inverse nano-emulsions consisting of 90 wt% toluene as the continuous phase and 10 wt% TDW as the dispersed phase, and diethylene triamine (DETA) and diphenylmethane-4,4′-diisocyanat (4,4′-MDI) as the amine and isocyanate monomers.

Aggregates were formed after adding the diisocyanate monomer 4,4′-MDI to systems containing 1 wt% and 5 wt% of anionic surfactant (sodium 1,4-bis(2-ethylhexoxy)-1,4-dioxobutane-2-sulfonate) (AOT), non-ionic surfactants sorbitane monooleate (SPAN80), or polyoxyethylene (2) stearyl ether (Brij 72). Aggregates were also obtained when 1 wt% of amphipathic emulsifier lecithin was used, as confirmed by scanning electron microscopy (SEM) analysis (Figure S1a). However, aggregates with some nano-capsules having incomplete polymerization were obtained when 1 wt% nonionic polymeric surfactant Agrimer AL22 was employed (Figure S1b), whereas nano-capsules with core-shell morphology were obtained with 1 wt% cetyl polyethyleneglycol/polypropyleneglycol-10/1 dimethicone (ABIL EM 90) (PU-1, Table S1), as confirmed by SEM and scanning transmission electron microscopy (STEM) analysis (Figure 2).

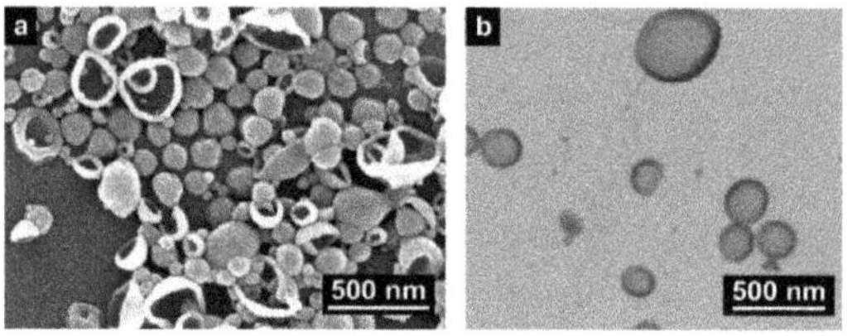

Figure 2. The morphology of PU NCs prepared using 1 wt% of ABIL EM 90 as the surfactant: (**a**) SEM and (**b**) STEM images.

The presence of the nano-capsules strongly indicates that polymeric surfactants are more compatible with the W/O system. This could be attributed to the fact that the polymeric surfactants produce several adsorption sites with negligible desorption from the emulsion interface. As shown by SEM, ABIL EM 90 provided better results than Agrimer AL22, indicating that ABIL EM 90 is more compatible in inverse nano-emulsion systems, presumably due to the enhancement of droplet stability and the strength of the interfacial film. Further optimizations, such as varying the surfactant percentage, the type and ratio of amine and isocyanate monomers, and varying the organic solvents, were carried out with ABIL EM 90 as the surfactant.

The stability of the W/O nano-emulsion as well as the resulting PU NCs depends not only on the type—it also depends on the amount of the applied surfactant. Therefore, different percentages (0.25%, 0.5%, 1%, 2%, 3%, 4%, and 5%) of ABIL EM 90 were employed. SEM analysis clearly indicated the formation of aggregates with less than 1% surfactant (Figure S2a,b). Interestingly, the size and polydispersity as well as the stability of the PU NCs were hardly affected by increasing the percentage of ABIL EM 90, however, at 3% and more, the PU NCs became more attached, as observed in SEM (Figure S2e–g). This shows that 1% (Figure S2c) and 2% (Figure S2d) of the surfactant afforded the best results. Because it is preferable to employ the minimum amount of surfactant, 1% ABIL EM 90 was chosen as the optimal condition and was used for additional optimization steps.

2.1.2. Variation of the Type and Percentage of the Continuous Organic Phase

Further optimization was carried out by examining the effect of the organic solvent on the stability and morphology of the PU NCs while utilizing the same composition mentioned above with 1 wt% ABIL EM 90, DETA (2.9 mmol), and 4,4′-MDI (2.9 mmol). Aggregates were formed with cyclohexane as the continuous phase (Figure S3a). However, in the presence of heptane, a mixture of PU particles, aggregates, and NCs was obtained (Figure S3b), whereas xylene provided a result similar to toluene (Figure S3c). These diverse morphologies were obtained presumably due to solubility differences regarding the isocyanate monomer in the aliphatic and aromatic continuous organic phases. In this regard, 4,4′-MDI has better solubility in toluene and xylene, compared with heptane and cyclohexane, hence, aggregates were formed in the aliphatic solvents. An additional parameter affecting the morphology is related to interfacial tension between the aqueous and organic phases. ABIL EM 90, at the specific applied concentration, seems to sufficiently lower the interfacial tension between toluene and water, rather than between cyclohexane or heptane and water, thus enhancing the stability of the nano-emulsion droplets and forming the desired nano-capsules.

To further examine the influence of the continuous phase on the formation and size distribution of the PU NCs, different percentages of a continuous phase (70%, 80% 85%, and 90%) consisting of toluene and ABIL EM 90 (1 wt%) were investigated. SEM analysis revealed the formation of polydispersed PU NCs systems with all percentages tested (Figure 3).

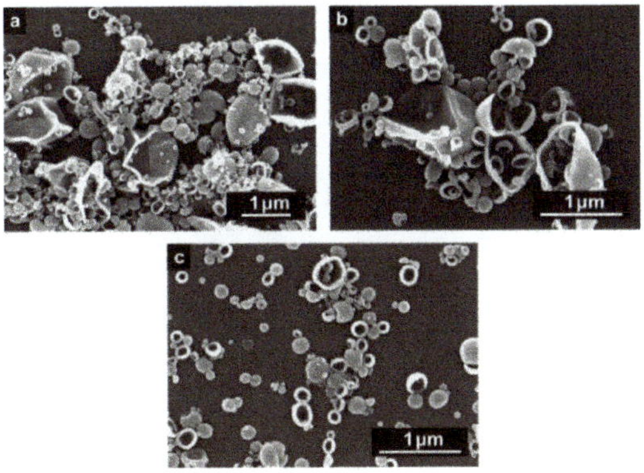

Figure 3. SEM images of PU NCs prepared using different continuous: aqueous phase ratios: (**a**) 70%:30%, (**b**) 80%:20 %, and (**c**) 85 %:15 %.

In addition, dynamic light scattering (DLS) studies revealed an increase in the average PU NCs size, along with a decrease in the percentage of the continuous phase and a simultaneous increase in the amount of the aqueous phase. When 70%, 80%, 85%, and 90% of the continuous phase were used, average sizes of 897 nm, 655 nm, 242 nm, and 270 nm were obtained, respectively (Figure 4).

2.1.3. Variation of the Type of the Polyurea (PU) Monomers

Initially, DETA and 4,4′-MDI were utilized as the amine and isocyanate monomers. An additional optimization step was conducted by varying the type and ratio of the amine and isocyanate monomers utilizing 1% of ABIL EM 90 and 10% of the aqueous phase containing poly(acrylamide-*co*-diallyldimethylammonium chloride (polyquaternium 7) as the lipophobe. This process appears to be essential, since the shell thickness and flexibility as well as the porosity and permeability of the PU NCs play a key role in the effectiveness and applicability of the system. Therefore, various amine monomers such DETA, ethylenediamine (EDA), 1,6-hexamethylenediamine (HMDA), and isocyanate monomers such as (PAPI 27, 4,4′-MDI, toluenediisocyanate (TDI), 1,6-hexamethylene diisocyanate (HDI), and 4,4′-methylenebis(cyclohexy isocyanate) (HMDI) were tested.

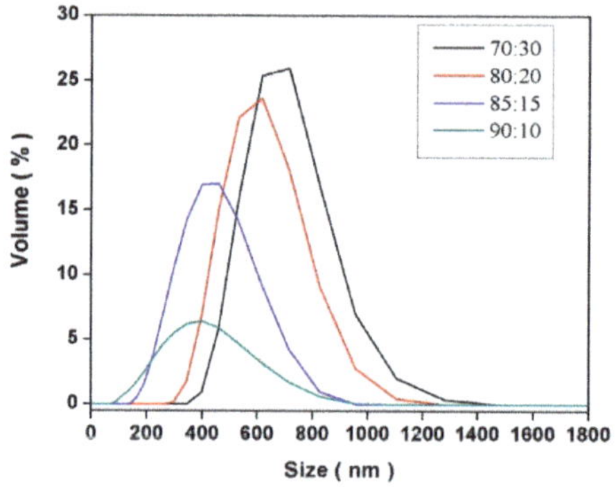

Figure 4. The influence of continuous: dispersed phase ratios of 70%:30%, 80%:20%, 85%:15%, and 90%:10% on the PU NCs size.

The combination of EDA with HMDI (PU-6), PAPI 27 (PU-5), 4,4′-MDI (PU-8), or TDI (PU-7), as well as DETA with TDI (PU-4), PAPI 27 (PU-2), or HMDI (PU-3) (Table S1) afforded either aggregates or an incomplete formation of the PU shell, as observed in SEM (Figure S4). Polydispersed PU NCs with an average size of 259 nm, as revealed by SEM, transmission electron microscopy (TEM), and DLS analyses (Figure 5), were obtained when 4,4′-MDI and a mixture of DETA and EDA were utilized as the PU monomers (PU-9, Table S1).

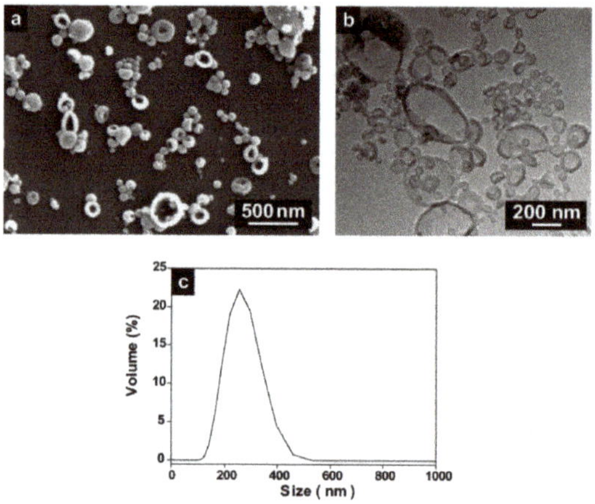

Figure 5. (**a**) SEM, (**b**) TEM, and (**c**) the size distribution of PU NCs prepared using 4,4′-MDI and a mixture of DETA and EDA as the monomers (PU-9).

2.1.4. The Influence of the Electrolyte

In order to suppress Ostwald ripening and obtain stable and non-aggregated capsules, an electrolyte is usually added to the aqueous phase. The influence of the electrolyte polyquaternium 7 on the size and morphology of the PU NCs was further studied using the PU-1 system. Varying the percentage of polyquaternium 7, namely, 0% (PU-10), 1% (PU-11), 2% (PU-12), and 3% (PU-13) (Table S1) had no appreciable effect on the morphology of PU NCs, as observed in SEM (Figure S5a–d). However, with 1% NaCl (PU-14, Table S1), aggregates and flattened capsules were obtained (Figure S5e). Additionally, DLS measurements revealed that the size distribution of PU-10 and PU-13 (Figure S6a,b) were similar to PU-1. These analyses clearly show that PU NCs are not affected by the electrolytes and can be formed even without utilizing them. An additional system with 80%:20% of continuous and aqueous phases respectively, without electrolyte was prepared (PU-15, Table S1). In this case, PU

NCs with a core shell structure were also formed, as confirmed by SEM and TEM (Figure 6a,b. DLS analysis, in agreement with SEM, revealed a polydispersed system with an average nano-capsule size of 365 nm (Figure 6c).

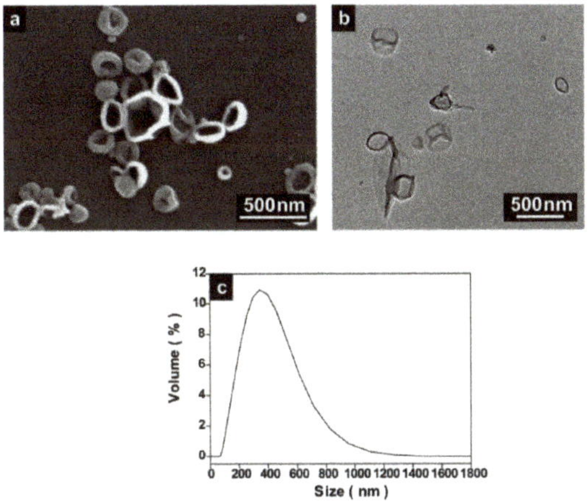

Figure 6. (**a**) SEM, (**b**) TEM, and (**c**) the size distribution of PU-15 prepared using 80%:20% of the continuous and dispersed phases, respectively.

The complete polymerization of the amine with the isocyanate monomers was revealed by Fourier transform infrared (FTIR) analysis (Figure 7). The presence of absorbance peaks above 3000 cm^{-1}, which correspond to the overlapping stretching vibrations of the N–H and –OH (from water) groups, and an absorption band at 1659 cm^{-1} assigned to C=O stretching vibrations, indicate the formation of polyurea [18]. In addition, the absence of the absorbance peak of the isocyanate group (–C=N=O) at 2260 cm^{-1} indicates the complete consumption of the isocyanate monomer. Furthermore, absorbance peaks at 1599 cm^{-1} and 1409 cm^{-1} can be attributed to C=C stretching vibrations of aromatic rings. The absorbance bands at 2852 and 2922 cm^{-1} are ascribed to sp3-hybridized C–H, and the peak at 3028 cm^{-1} is attributed to sp2 C–H stretching vibrations. In addition, absorbance bands at 1541 and 1095 cm^{-1} are ascribed to the secondary N–H bending and the C–N stretching vibrations, respectively.

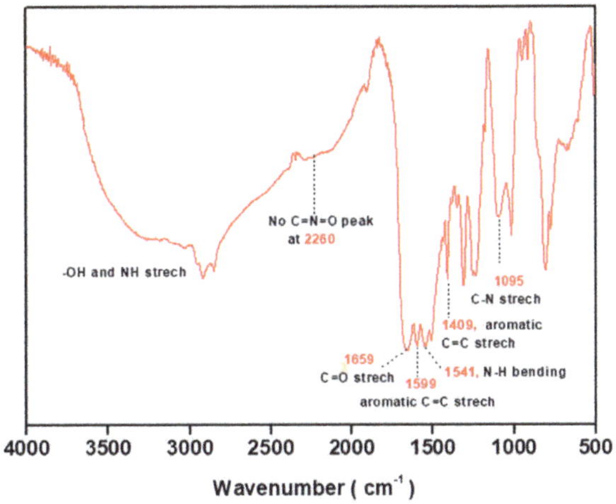

Figure 7. The FTIR spectrum of PU-15.

The formation of the PU NCs (PU-15) was further confirmed by solid state carbon nuclear magnetic resonance (^{13}C CP-MAS NMR) (Figure 8). Peaks at 21–48 ppm and 115–135 ppm are attributed to the aliphatic (from the DETA and EDA monomer) and aromatic (from 4,4′-MDI monomer) carbons of the PU shell. The peak at 156 ppm is ascribed to the carbonyl group (C=O) of the urea, indicating the formation of polyurea [18,30].

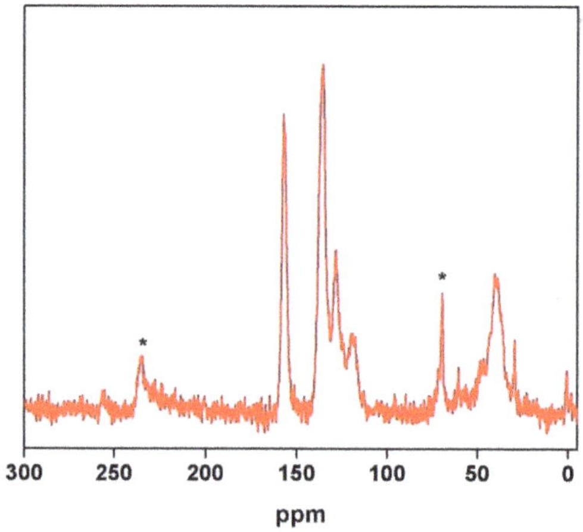

Figure 8. The ^{13}C CP-MAS NMR of PU NCs (PU-15); * indicates the spinning sideband.

2.2. Encapsulation of Ionic Liquid-Modified Magnetite Nanoparticles (MNPs-IL-C_4@PU NCs)

Magnetic polymer nanomaterials that combine the properties of organic and inorganic components have been fabricated for various applications in biomedical, environmental, sensing, drug delivery, and magnetic resonance imaging (MRI) [31–36].

Magnetic nanoparticles have also been applied in catalysis to facilitate the isolation and recovery of the nano-catalysts and catalyst nano-supports by applying an external magnetic field. Herein, we have prepared magnetic PU NCs in an inverse nano-emulsion (Figure 9a), which could have great potential in the above-mentioned fields.

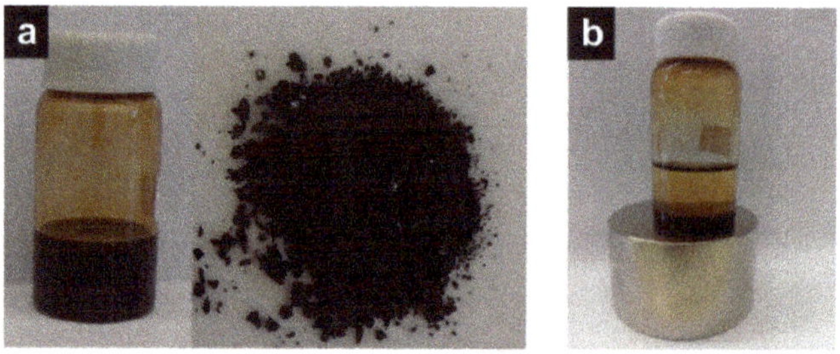

Figure 9. (**a**) MNPs-IL-C_4@PU NCs as suspension and powder and (**b**) a magnetic separation of MNPs-IL-C_4@PU NCs.

Incorporating MNPs within the PU NCs would impart superparamagnetic properties to the NCs, enabling them to be suitable for various applications such as magnetic nano-catalysts, which can be conveniently recovered under an external magnetic field (Figure 9b). We encapsulated different percentages (1–11 wt% with respect to the aqueous phase) of pre-prepared and stabilized MNPs with ionic liquid-based silane, 1-butyl-3-(3-(trimethoxysilyl)propyl)-1H- imidazol-3- chloride (IL-C_4) [25],

($MNPs-IL-C_4$) with an average size of 5–20 nm in the core of PU NCs using a composition employed for preparing the PU-15 system (Table 1).

Table 1. Composition and encapsulation of $MNPs-IL-C_4$ in PU-15 [a].

Entry	Amine Monomer	Isocyanate Monomer	$MNPs-IL-C_4$	Size (nm) [b]
PU-15	DETA/EDA	4,4′-MDI	0%	365.5
PU-15a	DETA/EDA	4,4′-MDI	1%	233.7
PU-15b	DETA/EDA	4,4′-MDI	2%	216.4
PU-15c	DETA/EDA	4,4′-MDI	3%	221.8
PU-15d	DETA/EDA	4,4′-MDI	5%	195.1
PU-15e	DETA/EDA	4,4′-MDI	7%	198.5
PU-15f	DETA/EDA	4,4′-MDI	9%	246.8
PU-15g	DETA/EDA	4,4′-MDI	11%	238.7

[a] Continuous phase (40 g, 80%) consisting of toluene (37.5 g) and ABIL EM 90 (2.5 g, 5 wt%). Dispersed phase (10 g, 20%) consisting of TDW (9.53 g), DETA (2.9 mmol), and EDA (2.9 mmol). Isocyanate (5.83 mmol) was dissolved in 10 g of total toluene and slowly added to the nano-emulsion system. [b] The average size was measured by dynamic light scattering (DLS).

The surface morphology of PU-15a-15g was probed by SEM (Figure 10) and TEM (Figure 11) analyses, which confirmed the formation of PU NCs with $MNPs-IL-C_4$ encapsulated in the core of the NCs. Since $MNPs-IL-C_4$ is hydrophilic, it does not dissolve in the non-polar continuous phase, therefore, non-encapsulated MNPs were not observed. Additionally, the size distribution of $MNPs-IL-C_4$@PU NCs decreased, compared with pure PU NCs, however, increasing the amount of MNPs had little effect on the polydispersity and average size distribution (Table 1, Figure S7). Moreover, FTIR analysis of PU-15a-15g exhibited similar results as pure PU NCs (PU-15), confirming the formation of polyurea shell (Figure S8).

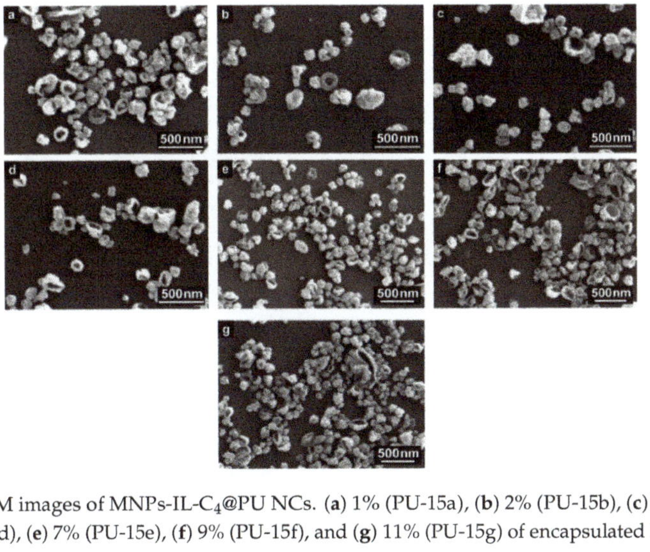

Figure 10. SEM images of $MNPs-IL-C_4$@PU NCs. (**a**) 1% (PU-15a), (**b**) 2% (PU-15b), (**c**) 3% (PU-15c), (**d**) 5% (PU-15d), (**e**) 7% (PU-15e), (**f**) 9% (PU-15f), and (**g**) 11% (PU-15g) of encapsulated $MNPs-IL-C_4$.

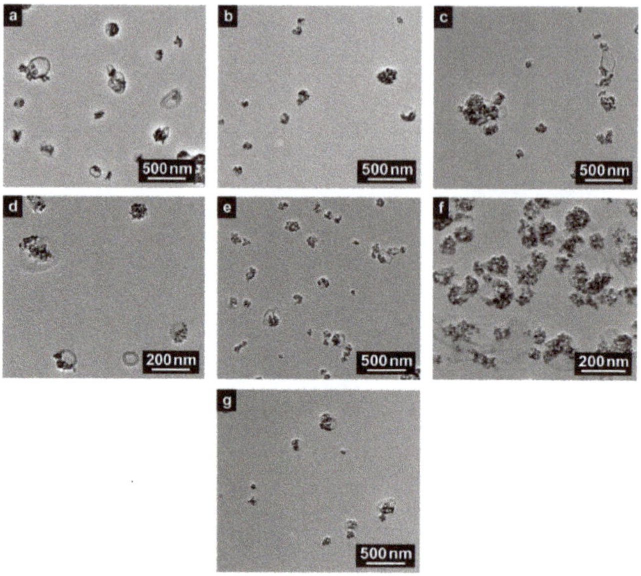

Figure 11. TEM images of MNPs-IL-C_4@PU NCs. (**a**) 1% (PU-15a), (**b**) 2% (PU-15b), (**c**) 3% (PU-15c), (**d**) 5% (PU-15d), (**e**) 7% (PU-15e), (**f**) 9% (PU-15f), and (**g**) 11% (PU-15g) of encapsulated MNPs-IL-C_4.

STEM/energy dispersive X-ray spectroscopy (EDS) and element mapping analyses were carried out in order to obtain more information about the composition and structure of the MNPs-IL-C_4@PU NCs. STEM/EDS analysis (Figure 12a) also confirmed the presence of an iron element content of the MNPs in the core of the PU NCs and showed the existence of an Si element attributed to the silane group of IL-C_4 supported on the MNPs and of ABIL EM 90.

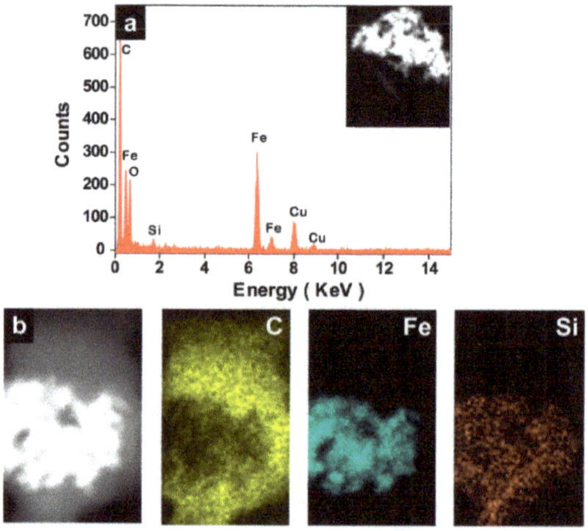

Figure 12. (**a**) STEM/ EDS and (**b**) EDS mapping analyses of MNPs-IL-C_4@PU NCs (PU-15b).

The elemental distribution of MNPs-IL-C_4@PU NCs was further probed by EDS mapping analysis (Figure 12b). In agreement with TEM, EDS mapping displayed the distribution of iron (Fe), indicating that the MNPs are located in the core of the PU NCs (Figure 12b, blue map). When comparing the distribution zones of C with the Si and Fe elements, it can be clearly seen that the C element (Figure 12b, yellow map) of the PU skeleton is distributed throughout all the areas of the NCs, whereas the Si element (Figure 12b, orange map), which is a component of silane IL-C_4 and ABIL EM 90, is more localized in the center.

The composition of the PU NCs, pure MNPs, and MNPs-IL-C_4@PU NCs was also probed by X-ray powder diffraction (XRD). The XRD pattern of PU NCs (PU-15) displayed a broad peak in the range 2θ = 10–30°, which is attributed to the amorphous polyurea. The XRD of PU-15a, PU-15d, and PU-15g revealed the presence of amorphous polyurea and showed the characteristic peaks of MNPs at 2θ = 18.1°, 30.2°, 35.6°, 43.2°, 53.6°, 57.2°, and 62.8° (Figure 13).

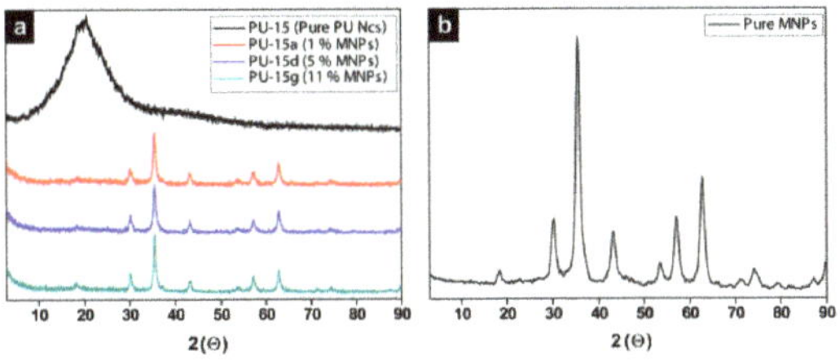

Figure 13. XRD pattern of (**a**) PU NCs (PU-15) and MNPs-IL-C_4@PU NCs and (**b**) pure MNPs.

The thermogravimetric analysis (TGA) of pure PU NCs and MNPs-IL-C_4@PU NCs (1–11%) over a temperature range of 25–950 °C under an inert atmosphere and at a heating rate of 10 °C/min revealed that pure PU NCs (PU-15) have two degradation steps with a total 94.7% weight loss (Figure 14). The first decomposition step was observed at 125–390 °C, which is attributed to the initial decomposition of the PU shell, toluene, and water [30,37]. The second step, which was at a temperature higher than 390 °C, is attributed to the additional decomposition of PU. The TGA curves of MNPs-IL-C_4@PU NCs revealed that when the 1, 2, 3, 5, 7, 9, or 11 wt% of MNPs per aqueous phase was added during the encapsulation process, the measured weight percentage of MNPs was 45.02%, 50.59%, 52.63%, 52.79%, 53.03%, 54.85%, or 55.64%, respectively. Moreover, TGA gives an indication of the stability of the PU NCs. It was clearly seen that all MNPs-IL-C_4@PU NCs systems exhibited three degradation steps with the initial decomposition temperature at ~220 °C. This indicates that the presence of MNPs-IL-C_4 increased the thermal stability of the PU NCs. The degradation steps at 220–500 °C and >500 °C are attributed to the decomposition of the PU shell and the IL group attached. Theoretically, pure PU should exhibit a 100% weight loss, since it is composed of pure organic material. However, PU-15 revealed the presence of 5.3% of non-decomposable material, which could be attributed to the inorganic surfactant ABIL-EM 90 and some species formed during the heating process.

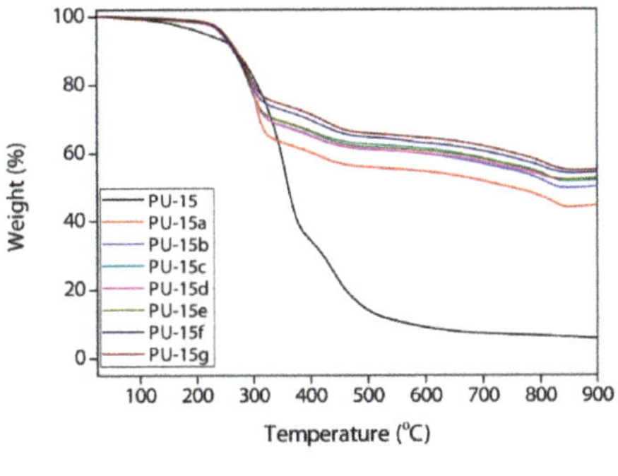

Figure 14. TGA analysis of pure PU NCs (PU-15) and MNPs-IL-C_4@PU NCs (1–11%).

3. Materials and Methods

Cetyl polyethyleneglycol/polypropyleneglycol-10/1 di-methicone (ABIL EM 90) was denoted by Sol-Gel Technologies (Ness Ziona, Israel); poly(1-ethenylpyrrolidin-2-one/hexadec-1-ene) (Agrimer

AL 22) and polymethylene-polyphenyl isocyanate (PAPI 27) were contributed by FMC Corporation (Ewing, NJ, USA). $FeCl_3 \times 6H_2O$, $FeCl_2 \times 4H_2O$, and ammonium hydroxide 25% were purchased from Acros Fischer Scientific through their distributor in Israel, Holland Moran LTD (Yehud, Israel). (3-chloropropyl) trimethoxysilane and 1-butyl imidazole were purchased from Sigma Aldrich (Rehovot, Israel). All amine and isocyanate monomers were purchased from Sigma-Aldrich or Acros Fischer.

Scanning electron microscopy (SEM) was utilized to determine the morphology of the PU NCs. The SEM analyses were carried out using a high-resolution scanning electron microscope (HR SEM) Sirion (FEI Company, Hillsboro, OR, USA) using a Schottky-type field emission source and a secondary electron detector. The images were scanned at a voltage of 5 kV. Transmission electron microscopy (TEM) and scanning transmission electron microscopy/energy dispersive X-ray spectroscopy (STEM/EDS) were performed with (S) TEM Tecnai F20 G2 (FEI Company, Hillsboro, OR, USA) operated at 200 kV. The Fourier transform infrared spectra (FTIR) were recorded at room temperature in transmission mode using a Perkin Elmer spectrometer 65 FTIR instrument (Waltham, MA, USA). Thermogravimetric analysis (TGA) was performed on a Mettler Toledo TG 50 analyzer (Greifensee, Switzerland). Measurements were carried out over a temperature range of 25–950 °C, at a heating rate of 10 °C/min under nitrogen. Dynamic light scattering (DLS) was utilized to determine the size distribution of the PU-NCs. These measurements were performed on a Nano Series instrument of model Nano-Zeta Sizer (Malvern Instruments, Worcestershire, United Kingdom) model ZEN3600. Powder X-ray diffraction (XRD) measurements were performed on a D8 Advance diffractometer (Bruker AXS, Karlsruhe, Germany) with a goniometer radius of 217.5 mm, a secondary graphite monochromator, with 2° Sollers slits, and a 0.2 mm receiving slit. Low-background quartz sample holders were carefully filled with the powder samples. XRD patterns within the range $2\theta = 1°$ to 90° were recorded at room temperature using CuKα radiation (λ = 1.5418 Å) with the following measurement conditions: a tube voltage of 40 kV, a tube current of 40 mA, step-scan mode with a step size of $2\theta = 0.02°$, and a counting time of 1 s/step. A solid-state ^{13}C NMR spectrum was recorded with a Bruker DRX-500 instrument (Rheinstetten, Germany).

3.1. *Synthesis of 1-Butyl-3-(3-(Trimethoxysilyl)Propyl)-1H-Imidazol-3-Cholride (IL-C_4)*

IL-C_4 was synthesized using procedures reported previously [25]. Briefly, 3-chloropropyltrimethoxysilane (14.2 g, 114.3 mmol) and 1-butylimidazole (22.73 g, 114.3 mmol) were stirred under an inert atmosphere at 120 °C for 24 h. The mixture was cooled to room temperature to obtain a yellow-orange viscous liquid (35.32 g, 96% yield).

3.2. *Preparation of Magnetite Nanoparticle-Supported IL-C_4 (MNPs-IL-C_4)*

MNPs-IL-C_4 was synthesized according to the procedure reported earlier [25].

3.3. *Preparation of Polyurea Nano-Capsules (PU NCs)*

The PU NCs were prepared in a typical procedure through interfacial polymerization in W/O nano-emulsion. The optimal PU NCs were prepared as follows: the continuous phase (organic phase, 40 g, 80%) consisted of toluene (37.5 g) and ABIL EM 90 (2.5 g, 5 wt%) was homogenized at 10,000 rpm for 30 s. A dispersed phase (aqueous phase, 10 g, 20%) consisted of triple-distilled water (TDW, 9.53 g), diethylene triamine (DETA, 2.9 mmol), and ethylenediamine (EDA, 2.9 mmol), which were then rapidly added during homogenization. The emulsification process was carried out for a further 1.5 min at 10,000 rpm, followed by sonication for 10 min using an ultrasonic cell disrupter with an output of 130 Watt and 20 KHz. Eventually, diphenylmethane-4,4′-diisocyanate (4,4′-MDI, 5.83 mmol), dissolved in 10 g total toluene, was slowly added to the nano-emulsion system during sonication. The mixture was then stirred for 3 h at room temperature. The resulting PU NCs were collected by centrifugation at 11,000 rpm for 15 min, washed two times with toluene, and finally re-dispersed in toluene to reach a 10 g suspension total.

3.4. Preparation of Magnetic PU NCs (MNPs-IL-C_4@PU NCs)

Magnetic PU NCs were prepared by encapsulation of MNPs-IL-C_4 in the PU NCs. The desired percentages of MNPs-IL-C_4 were dispersed in the aqueous phase and sonicated until all MNPs were fully dispersed. The encapsulation process was achieved by following the same procedure and using the same amount of surfactant and components as described in the preparation of PU NCs. Finally, the reaction mixture was mechanically stirred at room temperature for 3 h. The resulting magnetic PU NCs were separated by an external magnetic field and re-dispersed in toluene.

4. Conclusions

PU NCs with magnetite nanoparticles encapsulated within the aqueous core were prepared via interfacial polycondensation in inverse nano-emulsion.

Various parameters such as surfactant, solvents, monomers, and lipophobes were thoroughly examined in different ratios and compositions. The obtained systems were characterized for their morphology, chemical composition, thermal stability, size, and encapsulation efficiency. Among the examined parameters and conditions, it was found that the polymeric non-ionic surfactant with a low HLB value, ABIL EM 90, proved to be the best stabilizer for the inverse nano-emulsion and hence, for the established NCs. The desired PU NCs were obtained with DETA, EDA, and 4,4′-MDI as the PU monomers.

The morphology and size distribution of PU NCs were not affected by increasing the percentage of polyquaternium 7, indicating that the NCs can be formulated even without employing electrolytes as osmotic pressure agents. The encapsulation efficiency of PU NCs was examined by encapsulating up to 11% of MNPs-IL-C_4. The presence of the MNPs, regardless of the percentage, resulted in increased thermal stability of the PU NCs, as confirmed by TGA analysis. SEM and TEM analyses confirmed that the MNPs-IL-C_4 were not adsorbed on the capsules' shell but rather, were encapsulated in the aqueous core. Nonetheless, the average size distribution of the magnetic PU NCs decreased when compared with pure PU NCs. This could be due to the imidazolium group on the MNPs causing a reduction of Ostwald ripening. Owing to the facile synthesis and biocompatibility of PU, the magnetic properties of MNPs, and the expeditious magnetic separation of the system, the proposed magnetic PU NCs systems may be utilized in various applications such as catalysis, targeted delivery of hydrophilic drugs, and in other biomedical applications both in academia and industry.

Supplementary Materials: The following are available online. Figure S1–S5: SEM images, Table S1: System composition for preparing PU NCs, Figure S6–S7: Size distribution of pure PU NCs and MNPs-IL-C_4@PU NCs, Figure S8: FTIR spectra.

Author Contributions: S.N. prepared the nano-capsules. R.A.R and A.L.-Z. conceptualized the project, R.A.-R. supervised the experiments, R.A.-R and A.L.-Z. acquired the funding, S.N. prepared the original draft of the manuscript, and all coauthors contributed to writing the manuscript.

Funding: This research was funded by the Israel Ministry of Agriculture, [grant number 131-1595].

Acknowledgments: This work was supported by the Chief Scientist, the Israel Ministry of Agriculture grant # 131-1595. We are also grateful to the Ministry of Science, Technology, and Space for the fellowship of Suzana Natour. We thank Inna Popov and Vladimir Uvarov for helping with the TEM and XRD analysis. Suzana Natour thanks Rajashekharayya Sanguramath for productive discussions.

Conflicts of Interest: The authors declare no conflicts of interest. The funders had no role in the design of the study, nor in the collection, analyses, or interpretation of the data, in the writing of the manuscript, or in the decision to publish the results.

References

1. Fu, G.-D.; Li, G.L.; Neoh, K.G.; Kang, E.T. Hollow polymeric nanostructures—Synthesis, morphology and function. *Prog. Polym. Sci.* **2011**, *36*, 127–167. [CrossRef]
2. Guterres, S.S.; Alves, M.P.; Pohlmann, A.R. Polymeric nanoparticles, nanospheres and nanocapsules, for cutaneous applications. *Drug Target Insights* **2007**, *2*, 147–157. [CrossRef] [PubMed]

3. Pitaksuteepong, T.; Davies, N.M.; Tucker, I.G.; Rades, T. Factors influencing the entrapment of hydrophilic compounds in nanocapsules prepared by interfacial polymerisation of water-in-oil microemulsions. *Eur. J. Pharm. Biopharm.* **2002**, *53*, 335–342. [CrossRef]
4. Klapper, M.; Nenov, S.; Haschick, R.; Müller, K.; Müllen, K. Oil-in-oil emulsions: A unique tool for the formation of polymer nanoparticles. *Acc. Chem. Res.* **2008**, *41*, 1190–1201. [CrossRef] [PubMed]
5. Tiarks, F.; Landfester, K.; Antonietti, M. Preparation of polymeric nanocapsules by miniemulsion polymerization. *Langmuir* **2001**, *17*, 908–918. [CrossRef]
6. Johnsen, H.; Schmid, R.B. Preparation of polyurethane nanocapsules by miniemulsion polyaddition. *J. Microencapsul.* **2007**, *24*, 731–742. [CrossRef] [PubMed]
7. Mora-Huertas, C.E.; Garrigues, O.; Fessi, H.; Elaissari, A. Nanocapsules prepared via nanoprecipitation and emulsification–diffusion methods: Comparative study. *Eur. J. Pharm. Biopharm.* **2012**, *80*, 235–239. [CrossRef] [PubMed]
8. Landfester, K. Miniemulsions for nanoparticle synthesis. In *Colloid Chemistry II*; Antonietti, M., Ed.; Springer: Berlin/Heidelberg, Germany, 2003; pp. 75–123.
9. MacRitchie, F. Mechanism of interfacial polymerization. *Trans. Faraday Soc.* **1969**, *65*, 2503–2507. [CrossRef]
10. Gaudin, F.; Sintes-Zydowicz, N. Core–shell biocompatible polyurethane nanocapsules obtained by interfacial step polymerisation in miniemulsion. *Colloids Surf. A* **2008**, *331*, 133–142. [CrossRef]
11. Capek, I. On inverse miniemulsion polymerization of conventional water-soluble monomers. *Adv. Colloid. Interface Sci.* **2010**, *156*, 35–61. [CrossRef]
12. Cao, Z.; Ziener, U.; Landfester, K. Synthesis of narrowly size-distributed thermosensitive poly(*N*-isopropylacrylamide) nanocapsules in inverse miniemulsion. *Macromolecules* **2010**, *43*, 6353–6360. [CrossRef]
13. Cao, Z.; Ziener, U. Synthesis of nanostructured materials in inverse miniemulsions and their applications. *Nanoscale* **2013**, *5*, 10093–10107. [CrossRef] [PubMed]
14. Spernath, L.; Magdassi, S. Polyurea nanocapsules obtained from nano-emulsions prepared by the phase inversion temperature method. *Polym. Adv. Technol.* **2011**, *22*, 2469–2473. [CrossRef]
15. Landfester, K.; Willert, M.; Antonietti, M. Preparation of polymer particles in nonaqueous direct and inverse miniemulsions. *Macromolecules* **2000**, *33*, 2370–2376. [CrossRef]
16. Utama, R.H.; Stenzel, M.H.; Zetterlund, P.B. Inverse miniemulsion periphery RAFT polymerization: A convenient route to hollow polymeric nanoparticles with an aqueous core. *Macromolecules* **2013**, *46*, 2118–2127. [CrossRef]
17. Romio, A.P.; Rodrigues, H.H.; Peres, A.; Viegas, A.D.C.; Kobitskaya, E.; Ziener, U.; Landfester, K.; Sayer, C.; Araújo, P.H.H. Encapsulation of magnetic nickel nanoparticles via inverse miniemulsion polymerization. *J. Appl. Polym. Sci.* **2013**, *129*, 1426–1433. [CrossRef]
18. Dsouza, R.; Sriramulu, D.; Valiyaveettil, S. Topology and porosity modulation of polyurea films using interfacial polymerization. *RSC Adv.* **2016**, *6*, 24508–24517. [CrossRef]
19. Rosenbauer, E.-M.; Landfester, K.; Musyanovych, A. Surface-active monomer as a stabilizer for polyurea nanocapsules synthesized via interfacial polyaddition in inverse miniemulsion. *Langmuir* **2009**, *25*, 12084–12091. [CrossRef] [PubMed]
20. Crespy, D.; Stark, M.; Hoffmann-Richter, C.; Ziener, U.; Landfester, K. Polymeric nanoreactors for hydrophilic reagents synthesized by interfacial polycondensation on miniemulsion droplets. *Macromolecules* **2007**, *40*, 3122–3135. [CrossRef]
21. Wu, W.; He, Q.; Jiang, C. Magnetic iron oxide nanoparticles: Synthesis and surface functionalization strategies. *Nanoscale Res. Lett.* **2008**, *3*, 397–415. [CrossRef]
22. Zhu, Y.; Stubbs, L.P.; Ho, F.; Liu, R.; Ship, C.P.; Maguire, J.A.; Hosmane, N.S. Magnetic nanocomposites: A new perspective in catalysis. *ChemCatChem* **2010**, *2*, 365–374. [CrossRef]
23. Faraji, M.; Yamini, Y.; Rezaee, M. Magnetic nanoparticles: Synthesis, stabilization, functionalization, characterization, and applications. *J. Iran. Chem. Soc.* **2010**, *7*, 1–37. [CrossRef]
24. Baig, R.B.N.; Varma, R.S. Magnetically retrievable catalysts for organic synthesis. *Chem. Commun.* **2013**, *49*, 752–770. [CrossRef] [PubMed]
25. Natour, S.; Abu-Reziq, R. Immobilization of palladium catalyst on magnetically separable polyurea nanosupport. *RSC Adv.* **2014**, *4*, 48299–48309. [CrossRef]

26. Philippova, O.; Barabanova, A.; Molchanov, V.; Khokhlov, A. Magnetic polymer beads: Recent trends and developments in synthetic design and applications. *Eur. Polym. J.* **2011**, *47*, 542–559. [CrossRef]
27. El-Sherif, H.; El-Masry, M.; Emira, H.S. Magnetic polymer composite particles via in situ inverse miniemulsion polymerization process. *J. Macromol. Sci. Part A* **2010**, *47*, 1096–1103. [CrossRef]
28. Park, S.; Lee, Y.; Kim, Y.S.; Lee, H.M.; Kim, J.H.; Cheong, I.W.; Koh, W.-G. Magnetic nanoparticle-embedded PCM nanocapsules based on paraffin core and polyurea shell. *Colloids Surf. A* **2014**, *450*, 46–51. [CrossRef]
29. Kong, L.; Lu, X.; Jin, E.; Jiang, S.; Bian, X.; Zhang, W.; Wang, C. Constructing magnetic polyaniline/metal hybrid nanostructures using polyaniline/Fe_3O_4 composite hollow spheres as supports. *J. Solid State Chem.* **2009**, *182*, 2081–2087. [CrossRef]
30. Awad, W.H.; Wilkie, C.A. Investigation of the thermal degradation of polyurea: The effect of ammonium polyphosphate and expandable graphite. *Polymer* **2010**, *51*, 2277–2285. [CrossRef]
31. Schlegel, I.; Renz, P.; Simon, J.; Lieberwirth, I.; Pektor, S.; Bausbacher, N.; Miederer, M.; Mailänder, V.; Muñoz-Espí, R.; Crespy, D.; et al. Highly loaded semipermeable nanocapsules for magnetic resonance imaging. *Macromol. Biosci.* **2018**, *18*, 1700387. [CrossRef]
32. Pankhurst, Q.A.; Connolly, J.; Jones, S.K.; Dobson, J. Applications of magnetic nanoparticles in biomedicine. *J. Phys. D Appl. Phys.* **2003**, *36*, R167. [CrossRef]
33. Kalia, S.; Kango, S.; Kumar, A.; Haldorai, Y.; Kumari, B.; Kumar, R. Magnetic polymer nanocomposites for environmental and biomedical applications. *Colloid. Polym. Sci.* **2014**, *292*, 2025–2052. [CrossRef]
34. Alfadhel, A.; Li, B.; Kosel, J. Magnetic polymer nanocomposites for sensing applications. In Proceedings of the Sensors, Valencia, Spain, 2–5 November 2014; pp. 2066–2069.
35. Srinivasan, S.Y.; Paknikar, K.M.; Gajbhiye, V.; Bodas, D. Magneto-conducting core/shell nanoparticles for biomedical applications. *ChemNanoMat* **2018**, *4*, 151–164. [CrossRef]
36. Perera, A.S.; Zhang, S.; Homer-Vanniasinkam, S.; Coppens, M.-O.; Edirisinghe, M. Polymer–magnetic composite fibers for remote-controlled drug release. *ACS Appl. Mat. Interf.* **2018**, *10*, 15524–15531. [CrossRef] [PubMed]
37. Han, H.; Li, S.; Zhu, X.; Jiang, X.; Kong, X.Z. One step preparation of porous polyurea by reaction of toluene diisocyanate with water and its characterization. *RSC Adv.* **2014**, *4*, 33520–33529. [CrossRef]

Sample Availability: Samples of the compound 1-butyl-3-(3-(trimethoxysilyl)propyl)-1*H*-imidazol-3-cholride and MNPs-IL-C4@PU NCs are available from the authors.

molecules

MDPI

Article

Polydopamine-Based Surface Modification of ZnO Nanoparticles on Sericin/Polyvinyl Alcohol Composite Film for Antibacterial Application

Lisha Ai [1], Yejing Wang [1,2,*], Gang Tao [1], Ping Zhao [1,3], Ahmad Umar [4], Peng Wang [2] and Huawei He [1,3,*]

1 State Key Laboratory of Silkworm Genome Biology, Biological Science Research Center, Southwest University, Chongqing 400715, China; als123@email.swu.edu.cn (L.A.); taogang@email.swu.edu.cn (G.T.); zhaop@swu.edu.cn (P.Z.)

2 College of Biotechnology, Southwest University, Chongqing 400715, China; modelsums@email.swu.edu.cn

3 Chongqing Key Laboratory of Sericultural Science, Chongqing Engineering and Technology Research Center for Novel Silk Materials, Southwest University, Chongqing 400715, China

4 Department of Chemistry, College of Science and Arts and Promising Centre for Sensors and Electronics Devices, Najran University, P.O. Box: 1988, Najran 11001, Saudi Arabia; umahmad@nu.edu.sa

* Correspondence: yjwang@swu.edu.cn (Y.W.); hehuawei@swu.edu.cn (H.H.); Tel.: +86-23-6825-1575 (Y.W. & H.H.)

Academic Editors: Marinella Striccoli, Roberto Comparelli and Annamaria Panniello
Received: 21 December 2018; Accepted: 28 January 2019; Published: 30 January 2019

check for updates

Abstract: Silk sericin (SS) is a type of natural macromolecular protein with excellent hydrophilicity, biocompatibility and biodegradability, but also has very poor mechanical properties. To develop sericin-based wound dressings, we utilized polyvinyl alcohol (PVA) to reinforce the mechanical property of sericin by blending PVA and sericin, then modified zinc oxide nanoparticles (ZnO NPs) on SS/PVA film with the assistance of polydopamine (PDA) to endow SS/PVA film with antibacterial activity. Scanning electron microscopy, energy dispersive spectroscopy and X-ray powder diffraction demonstrated ZnO NPs were well grafted on PDA-SS/PVA film. Fourier transform infrared spectra suggested PDA coating and ZnONPs modification did not alter the structure of sericin and PVA. Water contact angle and swelling tests indicated the excellent hydrophilicity and swellability of ZnO NPs-PDA-SS/PVA composite film. Mass loss analysis showed ZnO NPs-PDA-SS/PVA film had excellent stability. The mechanical performance test suggested the improved tensile strength and elongation at break could meet the requirement of ZnO NPs-PDA-SS/PVA film in biomaterial applications. The antibacterial assay suggested the prepared ZnO NPs-PDA-SS/PVA composite film had a degree of antimicrobial activity against *Escherichia coli* and *Staphylococcus aureus*. The excellent hydrophilicity, swellability, stability, mechanical property and antibacterial activity greatly promote the possibility of ZnO NPs-PDA-SS/PVA composite film in antibacterial biomaterials application.

Keywords: ZnO nanoparticles; polydopamine; sericin; mechanical performance; antibacterial activity

1. Introduction

Wound dressing has a variety of functions, of which the first is to prevent continued bleeding through a certain degree of mechanical compression. In addition, wound dressing should have a certain degree of water absorption ability. It is also necessary for wound dressing to absorb the wound exudate in a timely manner, and at the same time to adequately maintain a moist environment to promote wound healing and reduce scarring formation [1]. Finally, some antibacterial properties are needed to prevent wound infection [2,3]. To meet the requirement of novel wound dressing, it is necessary to find natural materials with good hydrophilicity and biocompatibility that could be alternative candidates to traditional wound dressing. Natural silk produced by silkworm is composed of two parts: the outer part of silk collagen composition, known as sericin, and the inner part of core glial protein, known as silk fibroin [4,5]. Sericin is well known for having skin moisturizing properties as it has lots of amino acids with strong polar groups [6,7]. Sericin has huge potential in biomaterial applications due to its admirable hydrophilicity, biocompatibility and biodegradability [7]. More than 50,000 tons of sericin are discarded in waste water each year worldwide [8], causing an enormous waste of natural resources and environmental pollution. Hence, developing sericin-based biomaterials not only save resources, but also protect our environment. Heat treatment of silkworm cocoons at ambient or increased pressure is a preferred method of extracting sericin without introducing impurities [9,10]. Sericin is fragile in a relatively dry environment for its high content of random structures [11,12]. However, a certain degree of mechanical strength is required for a biomaterial. Therefore, reinforcing the mechanical properties of sericin is a huge challenge that must be resolved in expanding the application of sericin in biomaterials.

Recently, polyvinyl alcohol (PVA) has been extensively applied in biomaterials for its various excellent properties such as biocompatibility, hydrophilicity and mechanical performance [13]. PVA could effectively improve the mechanical properties of sericin and does not alter the other excellent properties of sericin [14,15]. In real life, the exposed skin wounds are susceptible to bacterial infections. Bacteria can enter into the deeper skin layers through wounds [16]. Therefore, it is required for sericin/PVA (SS/PVA) film to have the ability of preventing bacterial infection. Zinc oxide (ZnO) is a compound generally recognized as safe by the Food and Drug Administration of the United States of America (21CFR182.8991). ZnO has anti-viral, anti-bacterial and anti-fungal activities and minimal toxicity to humans [17–20]. ZnO nanoparticles (ZnO NPs) have a pronounced antibacterial ability with a high specific surface area to volume ratio. ZnO NPs can inhibit the growth of both Gram positive and Gram negative bacteria. The antibacterial activity of ZnO NPs is related to nanoparticles size. The smaller the size is, the greater the possibility for nanoparticles to have contact with the bacterial surface area. With the decrease of the particle size of ZnO NPs, its antibacterial properties increase. In addition, ZnO NPs have anti-corrosive and UV filtering properties. ZnO NP is easy to synthesize and has a type of green nanomaterials that possess good biocompatibility and biodegradability [21–23]. ZnO NPs have wide applications in food packaging and the cosmetics industry [22]. Hence, it is a good candidate to endow sericin with antibacterial activity for wound dressing. However, it is very difficult to graft ZnO NPs directly on the SS/PVA film.

Dopamine is an important neurohormone in the brain [24]. It has the catechins and amino groups with high affinity for metal ions, which can be used both as an ideal molecule covalent modification of starting point and a transition metal ions load anchor. These functional groups can pass on the basic properties of metal ions under the condition of having a strong restorative ability to further realize the emergence of a variety of mixed material [25]. With these benefits, polydopamine (PDA) is widely used in chemistry, biology, medicine, materials science and other fields. This performance promotes the deposition of metal on material surface [26]. PDA has excellent biocompatibility, biodegradability and hydrophilicity [27]. Taking advantage of these features, PDA could facilitate ZnO NPs being deposited on the SS/PVA composite film to endow it with antibacterial properties.

In this work, we first prepared SS/PVA film by blending sericin with PVA, soaked the composite film into PDA solution to coat PDA on its surface, and then immersed PDA-SS/PVA film into ZnO NPs solution to yield ZnO NPs-PDA-SS/PVA composite film. PDA was used to facilitate ZnO NPs being deposited on the SS/PVA composite film. Multiple techniques were performed by characterizing the composite films using scanning electron microscopy (SEM), Fourier transform infrared spectroscopy (FT-IR), X-ray diffractometry (XRD) and energy dispersive spectroscopy (EDS) to validate properties such as morphology and structure and also to prove the successful preparation of ZnO NPs on the film. Antimicrobial assays were carried out to investigate the antimicrobial activity of the composite film against *Escherichia coli* (*E. coli*) and *Staphylococcus aureus* (*S. aureus*).

2. Results and Discussion

ZnO NPs-PDA-SS/PVA composite film was prepared through the blending of sericin and PVA solution followed by PDA coating and ZnO NPs modification. The antibacterial activity of the composite film was tested to assess its usability in antibacterial applications such as wound dressing. The schematic diagram was shown in Figure 1.

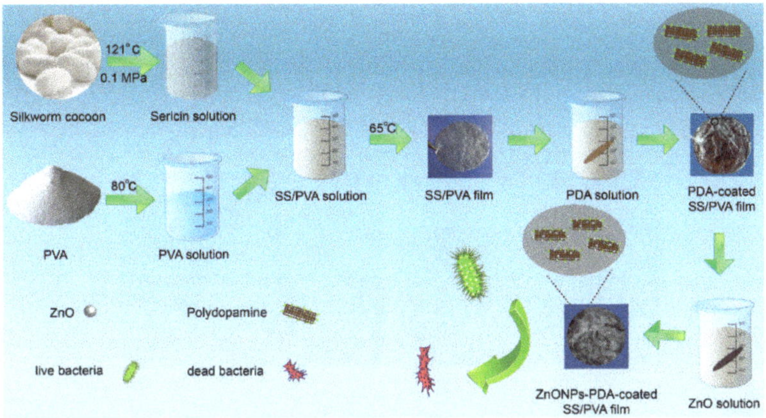

Figure 1. Schematic diagram of the preparation and antibacterial action of ZnO NPs-PDA-SS/PVA film.

2.1. SEM, EDS and XRD

SEM micrograph showed the surface morphology of the films. Figure 2a,d showed that SS/PVA film had a smooth surface, indicating the uniform blending of sericin and PVA. Figure 2b,e showed the surface morphology of PDA-SS/PVA film with a rough surface. PDA adsorption on the film is determined by monomer concentration, reaction time and temperature [14]. We carried out the experiment by constant stirring at 37 °C for 12 h in 2.0 mg/mL dopamine. PDA enhanced the interaction of sericin with ZnO NPs as a mediating agent. Figure 2c showed the surface morphology of ZnO NPs-PDA-SS/PVA composite film. The red arrows indicated ZnO NPs on the surface of PDA-SS/PVA film. Figure 2f showed the size distribution of ZnO NPs on the PDA-SS/PVA film. Most of ZnO NPs were in the size range of 60–120 nm, which implied its potential antibacterial activity.

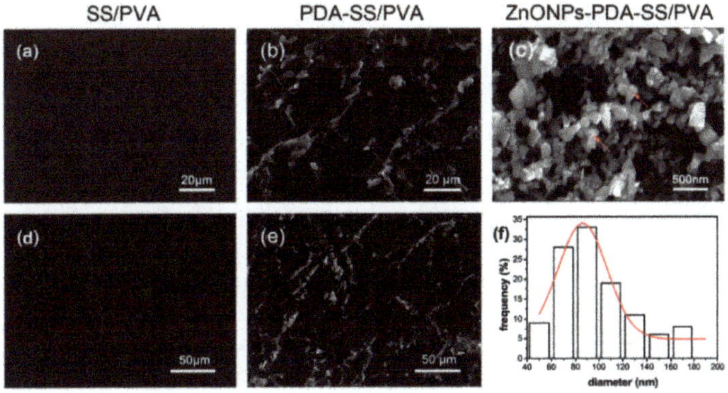

Figure 2. SEM of SS/PVA (**a**,**d**), PDA-SS/PVA (**b**,**e**) and ZnO NPs-PDA-SS/PVA films (**c**), respectively. The size distribution of ZnO NPs on the PDA-SS/PVA film (**f**).

To verify whether ZnO NPs were well grafted on the film, we performed energy-dispersive spectroscopy (EDS) analysis. The EDS spectrum of the selected area of the composite film (Figure 3a) showed the well-defined peaks of Zn and O, confirming the existence of ZnO on PDA-SS/PVA film (Figure 3c). This result proved that ZnO NPs were well grafted on the composite film. We also noticed variant shape and size of ZnO NPs on the surface of the film (Figure 3b).

XRD showed sericin has an abroad and shallow diffraction peak at around 2θ = 19.2°. A broad peak located at 2θ = 19.8° indicated the semicrystalline structure of sericin (Figure 3d). No typical diffraction peak of PDA was observed, indicating that PDA coating did not affect the crystal structure of SS/PVA composite film. The diffraction peaks observed at 2θ = 31.3°, 34.0°, 36.2°, 47.5°, 56,6°, 63.0°, and 68.9° corresponded to the crystal planes (100), (002), (101), (102), (110), (103), and (112), indicating the hexagonal formation of Wurtzite structure in ZnO nanoparticles [28].

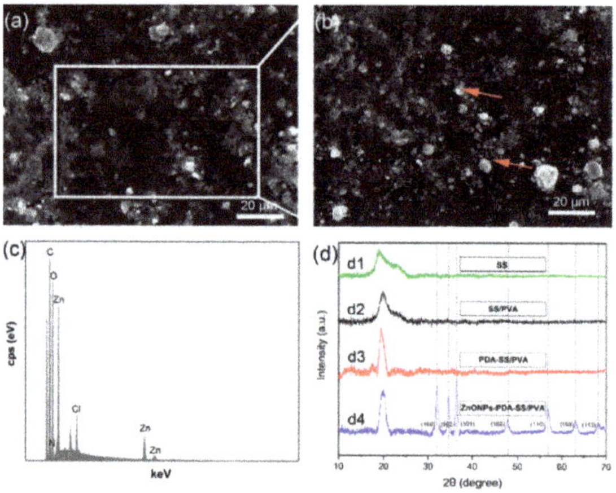

Figure 3. SEM characterization of ZnO NPs-PDA-SS/PVA film. (**a**,**b**) Field emission scanning electron microscope. ZnO NPs is indicated in red arrows. (**c**) A representative EDS spectrum of ZnO NPs-PDA-SS/PVA film. (**d**) The XRD profiles of sericin, SS/PVA, PDA-SS/PVA and ZnO NPs-PDA-SS/PVA films (d1–d4).

2.2. *FT-IR*

FT-IR was performed to characterize the chemical composition of sericin, SS/PVA, PDA-SS/PVA, ZnO NPs-PDA-SS/PVA films, as shown in Figure 4. The amide I, II and III bands of sericin existed in all spectra, corresponding to the characteristic peaks located at 1621 cm^{-1}, 1518 cm^{-1}, and 1240 cm^{-1} [29]. PVA has characteristic O-H stretching vibration peaks at 3276 cm^{-1} (Figure 4a) and 3269 cm^{-1} (Figure 4b) [30]. The peak of 1606 cm^{-1} indicates the C=C stretching and N-H deformation vibration of indoles or indoline structures in PDA molecule [31], suggesting that PDA was successfully grafted on the surface of SS/PVA composite film. The FT-IR spectrum of ZnO NPs-PDA-SS/PVA film was similar to that of PDA-SS/PVA film, suggesting ZnO NPs modification does not affect the amide peaks of sericin and the characteristic peaks of PVA and PDA.

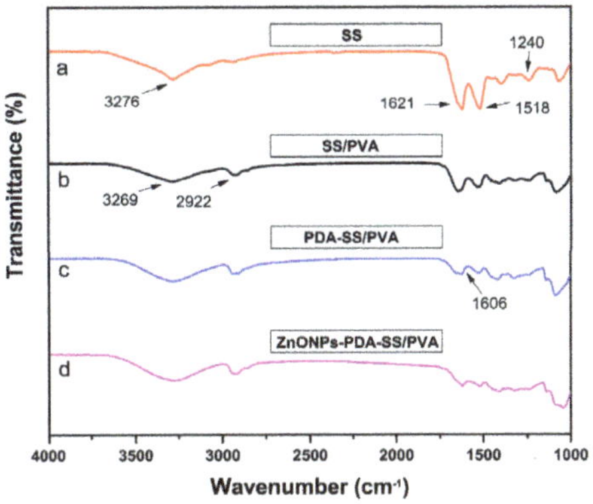

Figure 4. FT-IR spectra of the as prepared sericin and composite films (**a**–**d**).

2.3. Wettability and Swellability

Figure 5a–c showed the instantaneous water-absorption of SS/PVA, PDA-SS/PVA and ZnO NPs-PDA-SS/PVA composite films. The water contact angle of PDA-SS/PVA film was 29.8°, which was the smallest among all of the tested films, indicating that it has the best water absorption ability. PDA is an important factor to affect water absorption as it has excellent hydrophilicity [25]. The water contact angle of SS/PVA film was 56.7°, indicating the film is hydrophilic. After ZnO NPs modification, the water contact angle increased to 75.8°, but it was still hydrophilic. ZnO NPs modification covered the hydrophilic groups on the surface of PDA-SS/PVA film, which thus resulted in the decrease of water absorption property.

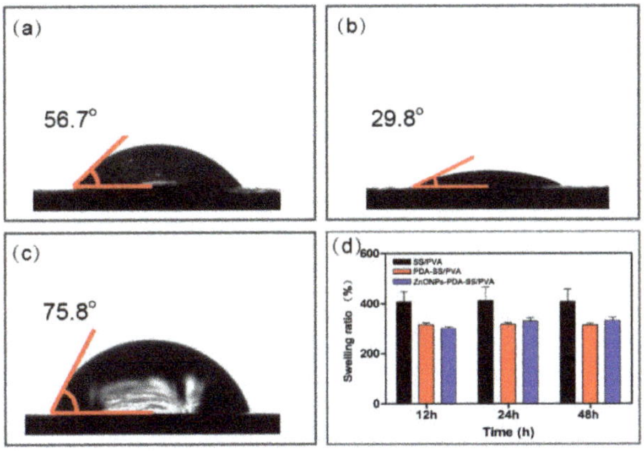

Figure 5. Water contact angle and swelling property of SS/PVA (**a**), PDA-SS/PVA (**b**), and ZnO NPs-PDA-SS/PVA films (**c**). Swelling ratio of these films (**d**) (n = 3 per group).

The swellability of SS/PVA, PDA-SS/PVA and ZnO NPs-PDA-SS/PVA composite films is shown in Figure 5d. All the composite films with a dimension of 1 cm × 1 cm (length × width) were immersed into PBS buffer (pH 7.4) for 12 h, 24 h and 48 h, respectively. The swelling ratio of SS/PVA increased about 4 times, indicating SS/PVA film has an excellent swellability. The swelling ratio of PDA-SS/PVA and ZnO NPs-PDA-SS/PVA films were about 3 times of the control, suggesting these films have excellent swellability. The swelling ratio did not change over time. This may be explained by the fact that the sites of water molecule binding have been completely saturated in a short time. The excellent hydrophilicity and swellability indicated the potential of ZnO NPs-PDA-SS/PVA in wound dressing applications.

2.4. Mechanical Property

The mechanical properties of SS/PVA, PDA-SS/PVA and ZnO NPs-PDA-SS/PVA films were shown in Figure 6. SS/PVA had the highest tensile strength among all of the tested films, while PDA-SS/PVA film was the lowest. ZnO-PDA-SS/PVA film had a tensile strength of about 8 MPa, which is suitable for application of wound dressings. The elongation at break is an indicator of the material's flexibility [32]. The results showed that PDA coating resulted in the increase of the elongation at break of SS/PVA film, while ZnO NPs modification reduced the elongation at break of PDA/PVA film (Figure 6b). PDA may interact with sericin/PVA through its amide and hydroxyl groups to form an extensive molecular network, and thus improved the elongation at break of SS/PVA film and decreased the tensile strength of SS/PVA film. ZnO NPs likely disrupted the molecular interaction between PDA and sericin/PVA partially, which resulted in the decrease of the elongation at break and the increase of the tensile strength. The elongation at break of these films ranged from 50% to 160%. The data showed ZnO NPs-PDA-SS/PVA film meets the requirements of wound dressing materials.

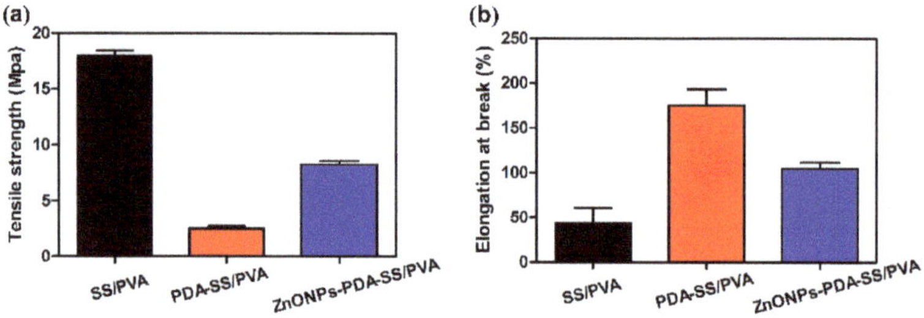

Figure 6. Mechanical properties of the films: (**a**) tensile strength and (**b**) elongation at break ($n = 3$ per group).

2.5. Mass Loss Analysis

Sustained stability is one of the characteristics that a wound dressing is supposed to have. Here, we analyzed the mass loss of ZnO NPs-PDA-SS/PVA film under pH 4.0, 7.4 and 10.0 conditions to assess its stability. The cumulative mass loss of the composite film increased over time. Under pH 10.0, the mass loss of the film occurred faster than that under pH 4.0 and 7.4 conditions (Figure 7). This may be that sericin contains a number of acidic amino acids and has an isoelectric point of 3.8 [33]. While in an alkaline environment, sericin more easily reacts and can be more easily hydrolyzed [34]. The result showed that ZnO NPs-PDA-SS/PVA composite film has good stability.

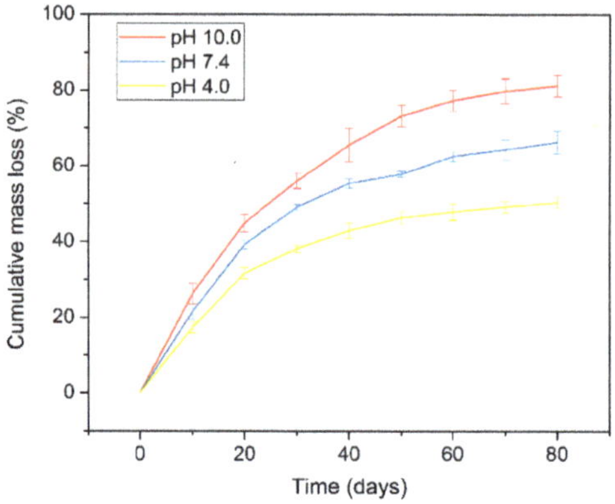

Figure 7. Mass loss of ZnO NPs-PDA-SS/PVA films under different pH conditions.

2.6. Antibacterial Property

The antibacterial property of ZnO NPs-PDA-SS/PVA, PDA-SS/PVA and SS/PVA films were analyzed against Gram-negative bacteria (*E. coli*) and Gram-positive bacteria (*S. aureus*), respectively. As shown in Figure 8, while compared to the control, the colonies number did not show a significant difference in the presence of SS/PVA and PDA-SS/PVA films. However, in the presence of ZnO NPs-PDA-SS/PVA film, the colonies number was much less than that of the control, indicating the good antibacterial property of the composite film against *E. coli* and *S. aureus*. The antibacterial activity of ZnONPs comes from three aspects: (a) Reactive oxygen species are produced from ZnO NPs, which destroy cell membrane, thus causing leakage of cytoplasmic contents, DNA damage, and cell death [35]; (b) Zinc ions from ZnO NPs can penetrate cell membrane to inhibit bacterial metabolic activity [32,36]; (c) The intracellular accumulation of ZnO NPs can destroy the cell wall of bacteria and affect DNA replication, leading to the death of bacteria [18].

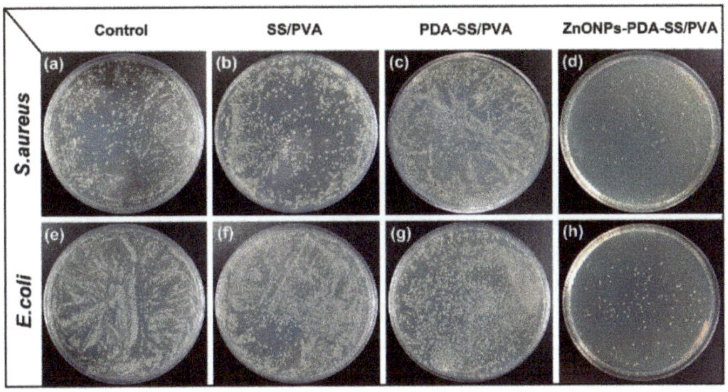

Figure 8. Antibacterial activity analysis of SS/PVA, PDA-SS/PVA and ZnO NPs-PDA-SS/PVA films against *E. coli* (**a**–**d**) and *S. aureus* (**e**–**h**).

2.7. Bacterial Growth Assay

To further confirm the bacteriostasis of ZnO NPs-PDA-SS/PVA film, the bacterial growth curve in the presence of the composite film was presented by measuring bacterial OD_{600}, as shown in Figure 9. The bacterial growth profile was very similar between SS/PVA, PDA-SS/PVA films and the control. However, the bacterial growth was significantly retarded in the presence of ZnO NPs-PDA-SS/PVA film, suggesting ZnO NPs-PDA-SS/PVA film has a certain inhibitory effect on the growth of *E. coli* and *S. aureus*. This result was consistent with that of the colony counting method.

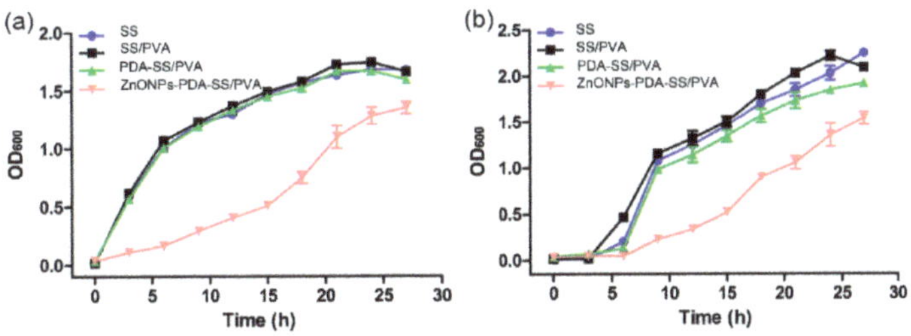

Figure 9. Bacterial growth curves of *E. coli* (**a**) and *S. aureus* (**b**) in the presence of different films.

3. Materials and Methods

3.1. Materials

Silkworm cocoons were supplied by the State Key Laboratory of Silkworm Genome Biology, Southwest University, Chongqing, China. PVA and dopamine hydrochloride were purchased from Aladdin Corp (Shanghai, China). ZnO NPs (99.9%) was purchased from Guohua reagent (Shanghai, China). All other chemicals were of analytical grade and were directly used.

3.2. Fabrication of ZnO NPs-PDA-SS/PVA Film

Sericin was obtained from silkworm cocoons by means of autoclave at 121 °C for 30 min [37,38]. SS/PVA film was prepared as per our previous procedure [37,38]. Briefly, sericin solution (4%, w/t) and PVA solution (5%, w/t) were well mixed, and then dried at 65 °C in the Petri dishes to become SS/PVA film. SS/PVA film was immersed directly into fresh dopamine solutions (2.0 mg/mL, pH 8.5) at 37 °C for 12 h with continuous stirring. Then the PDA coated SS/PVA film was taken out and washed with MilliQ water to remove extra PDA. Furthermore, PDA-SS/PVA film was soaked in 61.7 mM ZnO NPs at 25 °C for 12 h. Finally, the ZnO NPs-PDA-SS/PVA composite film was produced after repetitive washing and was dried at 25 °C.

3.3. SEM, XRD, FT-IR and Mechanical Test

The surface morphology of ZnO NPs-PDA-SS/PVA, PDA-SS/PVA and SS/PVA films were imaged on a JSM-5610LV scanning electron microscopy (Tokyo, Japan) with a working voltage of 25 kV. The crystalline structure of the composite film was examined by PANalytical X'Pert, a powder X-ray diffraction system (Almelo, The Netherlands) over Bragg angles ranging from 10° to 80°. FT-IR spectra of the composite film were characterized on Thermo-Fisher Nicolet iz10 IR microscope (Framingham, MA, USA) over the wavenumber of 4000–400 cm^{-1}.

SS/PVA, PDA-SS/PVA, ZnO NPs-PDA-SS/PVA films were prepared with a dimension of 4 cm × 1 cm (length × width). These films' tensile properties were tested on a universal AG-X-plus testing machine (Shimadzu, Kyoto, Japan) equipped with a 1000 N loading cell. The crosshead speed was 5 mm/min. The thickness of the film was measured by SEM. To reveal the stress-strain relation, the recorded data was transformed into real stress (σ) and strain (ε) [39].

3.4. Hydrophilicity and Swellability

The hydrophilicity of ZnO NPs-PDA-SS/PVA, PDA-SS/PVA and SS/PVA were analyzed at 25 °C on a Krüss DSA100 system (Hamburg, Germany) via sessile drop contact angle. At five different positions, the process of water absorption was recorded by dispensing the water droplet on the surface of the sample.

The swelling of the composite films in phosphate buffer (PBS, pH 7.4) was measured by gravimetric method with a minor modification. The original films were weighed as W_1. Then, the swelling films were immersed into phosphate buffer at 37 °C, then taken out at separate time intervals and gently removed by filter paper. The mass of swollen composite films was marked as W_2. The swelling ratio (S) was calculated by the following equation:

$$S\ (\%) = (W_2 - W_1) \times 100\%/W_1 \tag{1}$$

The same operation was performed at least three times under the same condition. The swellability of the film was presented by average data.

3.5. Mass Loss Analysis

A mass losing ratio analysis was conducted to analyze the stability of the composite films. First, the films were prepared with a dimension of 3 cm × 3 cm (length × width), and then soaked into

PBS with different values (pH 4.0, 7.4, 10.0) at 25 °C, respectively. At given time intervals, the films were carefully taken out, washed, and weighed. The mass of the film before and after the treatment were recorded as W_3 and W_4, respectively. The test was repeated five times. The mass losing ratio (R) was calculated using the following equation:

$$R\ (\%) = (W_3 - W_4) \times 100\% / W_3 \quad (2)$$

3.6. Antibacterial Test

The antibacterial activity of the composite film was assessed against gram-positive bacteria *S. aureus* and gram-negative bacteria *E. coli* as Pal's protocol [40]. First, the bacteria were cultured in 10 mL Luria-Bertani (LB) medium (pH 7.4) under a constant vibrating velocity of 2.8× *g* at 37 °C overnight. Then the bacteria suspension was diluted with LB medium to an optical density at 600 nm (OD_{600}) of 0.02, which was used as the control. The diluted bacteria suspension was cultured at 37 °C for 30 h in the presence of sterile ZnO NPs-PDA-SS/PVA, PDA-SS/PVA, SS/PVA films (1 cm × 1 cm), respectively.

The bacteria were cultured in 12-well plate in the presence of sterile composite films at 37 °C, and were collected at different time intervals to measure OD_{600}. After 3 h, the bacterial suspension of each experiment was diluted for several times and then cultured on LB plate at 37 °C overnight. The antibacterial activity of the film was evaluated by counting the number of colonies on the plate. The assay was performed in triplicate for each independent experiment.

4. Conclusions

In this work, we prepared SS/PVA film with enhanced mechanical performance through blending sericin and PVA, and then grafted ZnO NPs on SS/PVA film via PDA to yield ZnO NPs-PDA-SS/PVA composite film. SEM confirmed the spherical shape ZnO NPs on PDA-SS/PVA film. The elemental composition and chemical composition of ZnO NPs-PDA-SS/PVA film were confirmed by EDS and FT-IR. The crystal planes of ZnO NPs were determined by XRD. Water contact angle and swelling analysis indicated the excellent hydrophilicity and swellability of ZnO NPs-PDA-SS/PVA film. The mechanical test validated the improved mechanical performance of the composite film. Mass loss analysis showed good stability of ZnO NPs-PDA-SS/PVA film under different pH conditions. ZnO NPs-PDA-SS/PVA film exhibited a certain degree of antibacterial effect against *E. coli* and *S. aureus*. The improved mechanical performance and antibacterial activity will greatly promote the application of ZnO NPs-PDA-SS/PVA composite film in antimicrobial biomaterials such as wound dressing.

Author Contributions: L.A., Y.W. and H.H. conceived and designed the experiments; L.A., P.W. performed the experiments; L.A., G.T. and Y.W. analyzed the data; P.Z. contributed reagents/materials/analysis tools; L.A. and Y.W. wrote the draft; Y.W. and H.H. supervised the research; A.U. and H.H. revised the manuscript.

Funding: This work was supported by the National Natural Science Foundation of China (31572465), the State Key Program of the National Natural Science of China (31530071), Fundamental Research Funds for the Central Universities (XDJK2018B010, XDJK2018C063), the Graduate Research and Innovation Project of Chongqing (CYS18123, CYB17069) and the Open Project Program of Chongqing Engineering and Technology Research Center for Novel Silk Materials (silkgczx2016003).

Conflicts of Interest: The authors declare no conflict of interest. The founding sponsors had no role in the design of the study; in the collection, analyses, or interpretation of data; in the writing of the manuscript, and in the decision to publish the results.

References

1. Pan, H.; Fan, D.; Cao, W.; Zhu, C.; Duan, Z.; Fu, R.; Li, X.; Ma, X. Preparation and Characterization of Breathable Hemostatic Hydrogel Dressings and Determination of Their Effects on Full-Thickness Defects. *Polymers* **2017**, *9*, 727. [CrossRef]
2. Al-Omair, M. Synthesis of Antibacterial Silver–Poly(ε-caprolactone)-Methacrylic Acid Graft Copolymer Nanofibers and Their Evaluation as Potential Wound Dressing. *Polymers* **2015**, *7*, 1464–1475. [CrossRef]

3. Khil, M.S.; Cha, D.I.; Kim, H.Y.; Kim, I.S.; Bhattarai, N. Electrospun nanofibrous polyurethane membrane as wound dressing. *J. Biomed. Mater. Res. B* **2003**, *67*, 675–679. [CrossRef] [PubMed]
4. Kato, N.; Sato, S.; Yamanaka, A.; Yamada, H.; FUWA, N.; NOMURA, M. Silk protein, sericin, inhibits lipid peroxidation and tyrosinase activity. *Biosci. Biotechnol. Biochem.* **1998**, *62*, 145–147. [CrossRef]
5. He, H.; Tao, G.; Wang, Y.; Cai, R.; Guo, P.; Chen, L.; Zuo, H.; Zhao, P.; Xia, Q. In situ green synthesis and characterization of sericin-silver nanoparticle composite with effective antibacterial activity and good biocompatibility. *Mater. Sci. Eng. C* **2017**, *80*, 509–516. [CrossRef] [PubMed]
6. Kundu, B.; Rajkhowa, R.; Kundu, S.C.; Wang, X. Silk fibroin biomaterials for tissue regenerations. *Adv. Drug Deliv. Rev.* **2013**, *65*, 457–470. [CrossRef] [PubMed]
7. Liu, L.; Cai, R.; Wang, Y.; Tao, G.; Ai, L.; Wang, P.; Yang, M.; Zuo, H.; Zhao, P.; Shen, H. Preparation and characterization of AgNPs in situ synthesis on polyelectrolyte membrane coated sericin/agar film for antimicrobial applications. *Materials* **2018**, *11*, 1205. [CrossRef]
8. Lamboni, L.; Gauthier, M.; Yang, G.; Wang, Q. Silk sericin: A versatile material for tissue engineering and drug delivery. *Biotechnol. Adv.* **2015**, *33*, 1855–1867. [CrossRef]
9. Aramwit, P.; Siritientong, T.; Srichana, T. Potential applications of silk sericin, a natural protein from textile industry by-products. *Waste Manag. Res.* **2012**, *30*, 217–224. [CrossRef]
10. Liu, L.; Cai, R.; Wang, Y.; Tao, G.; Ai, L.; Wang, P.; Yang, M.; Zuo, H.; Zhao, P.; He, H. Polydopamine-Assisted Silver Nanoparticle Self-Assembly on Sericin/Agar Film for Potential Wound Dressing Application. *Int. J. Mol. Sci.* **2018**, *19*, 2875. [CrossRef]
11. Nayak, S.; Talukdar, S.; Kundu, S.C. Potential of 2D crosslinked sericin membranes with improved biostability for skin tissue engineering. *Cell Tissue Res.* **2012**, *347*, 783–794. [CrossRef] [PubMed]
12. He, H.; Cai, R.; Wang, Y.; Tao, G.; Guo, P.; Zuo, H.; Chen, L.; Liu, X.; Zhao, P.; Xia, Q. Preparation and characterization of silk sericin/PVA blend film with silver nanoparticles for potential antimicrobial application. *Int. J. Biol. Macromol.* **2017**, *104*, 457–464. [CrossRef] [PubMed]
13. Kim, H.; Lee, Y.; Kim, Y.; Hwang, Y.; Hwang, N. Biomimetically Reinforced Polyvinyl Alcohol-Based Hybrid Scaffolds for Cartilage Tissue Engineering. *Polymers* **2017**, *9*, 655. [CrossRef]
14. Cai, R.; Tao, G.; He, H.; Song, K.; Zuo, H.; Jiang, W.; Wang, Y. One-step synthesis of silver nanoparticles on polydopamine-coated sericin/polyvinyl alcohol composite films for potential antimicrobial applications. *Molecules* **2017**, *22*, 721. [CrossRef] [PubMed]
15. Tao, G.; Cai, R.; Wang, Y.; Song, K.; Guo, P.; Zhao, P.; Zuo, H.; He, H. Biosynthesis and characterization of AgNPs–silk/PVA film for potential packaging application. *Materials* **2017**, *10*, 667. [CrossRef] [PubMed]
16. Buja, A.; Zampieron, A.; Cavalet, S.; Chiffi, D.; Sandonà, P.; Vinelli, A.; Baldovin, T.; Baldo, V. An update review on risk factors and scales for prediction of deep sternal wound infections. *Int. Wound J.* **2012**, *9*, 372–386. [CrossRef] [PubMed]
17. Fernando, S.; Gunasekara, T.; Holton, J. Antimicrobial Nanoparticles: Applications and mechanisms of action. *Sri Lankan J. Infect. Dis.* **2018**, 8. [CrossRef]
18. Zhang, L.; Jiang, Y.; Ding, Y.; Povey, M.; York, D. Investigation into the antibacterial behaviour of suspensions of ZnO nanoparticles (ZnO nanofluids). *J. Nanopart Res.* **2007**, *9*, 479–489. [CrossRef]
19. Wang, Y.W.; Cao, A.; Jiang, Y.; Zhang, X.; Liu, J.H.; Liu, Y.; Wang, H. Superior antibacterial activity of zinc oxide/graphene oxide composites originating from high zinc concentration localized around bacteria. *ACS Appl. Mater. Interfaces* **2014**, *6*, 2791–2798. [CrossRef]
20. Lipovsky, A.; Nitzan, Y.; Gedanken, A.; Lubart, R. Antifungal activity of ZnO nanoparticles—The role of ROS mediated cell injury. *Nanotechnology* **2011**, *22*, 105101. [CrossRef]
21. Reddy, K.M.; Feris, K.; Bell, J.; Wingett, D.G.; Hanley, C.; Punnoose, A. Selective toxicity of zinc oxide nanoparticles to prokaryotic and eukaryotic systems. *Appl. Phys. Lett.* **2007**, *90*, 585. [CrossRef] [PubMed]
22. Dayakar, T.; Rao, K.V.; Bikshalu, K.; Rajendar, V.; Park, S.H. Novel synthesis and structural analysis of zinc oxide nanoparticles for the non enzymatic glucose biosensor. *Mater. Sci. Eng. C* **2017**, *75*, 1472–1479. [CrossRef] [PubMed]
23. Marra, A.; Rollo, G.; Cimmino, S.; Silvestre, C. Assessment on the Effects of ZnO and Coated ZnO Particles on iPP and PLA Properties for Application in Food Packaging. *Coatings* **2017**, *7*, 29. [CrossRef]
24. Zhang, A.; Neumeyer, J.L.; Baldessarini, R.J. Recent progress in development of dopamine receptor subtype-selective agents: Potential therapeutics for neurological and psychiatric disorders. *Chem. Rev.* **2007**, *107*, 274–302. [CrossRef] [PubMed]

25. Liu, Y.; Ai, K.; Lu, L. Polydopamine and its derivative materials: Synthesis and promising applications in energy, environmental, and biomedical fields. *Chem. Rev.* **2014**, *114*, 5057–5115. [CrossRef] [PubMed]
26. Mondin, G.; Wisser, F.M.; Leifert, A.; Mohamed-Noriega, N.; Grothe, J.; Dorfler, S.; Kaskel, S. Metal deposition by electroless plating on polydopamine functionalized micro- and nanoparticles. *J. Colloid Interface Sci.* **2013**, *411*, 187–193. [CrossRef]
27. Kang, S.M.; You, I.; Cho, W.K.; Shon, H.K.; Lee, T.G.; Choi, I.S.; Karp, J.M.; Lee, H. One-step modification of superhydrophobic surfaces by a mussel-inspired polymer coating. *Angew. Chem. Int. Ed.* **2010**, *49*, 9401–9404. [CrossRef]
28. Bagheri, M.; Rabieh, S. Preparation and characterization of cellulose-ZnO nanocomposite based on ionic liquid ([C_4 mim] Cl). *Cellulose* **2013**, *20*, 699–705. [CrossRef]
29. Zhang, X.; Wyeth, P. Using FTIR spectroscopy to detect sericin on historic silk. *SCI China Chem.* **2010**, *53*, 626–631. [CrossRef]
30. Kiro, A.; Bajpai, J.; Bajpai, A. Designing of silk and ZnO based antibacterial and noncytotoxic bionanocomposite films and study of their mechanical and UV absorption behavior. *J. Mech. Behav. Biomed. Mater.* **2017**, *65*, 281–294. [CrossRef]
31. Dreyer, D.R.; Miller, D.J.; Freeman, B.D.; Paul, D.R.; Bielawski, C.W. Elucidating the structure of poly (dopamine). *Langmuir* **2012**, *28*, 6428–6435. [CrossRef] [PubMed]
32. Li, Y.; Zhang, W.; Niu, J.; Chen, Y. Mechanism of photogenerated reactive oxygen species and correlation with the antibacterial properties of engineered metal-oxide nanoparticles. *ACS Nano* **2012**, *6*, 5164–5173. [CrossRef] [PubMed]
33. Wang, Z.; Zhang, Y.; Zhang, J.; Huang, L.; Liu, J.; Li, Y.; Zhang, G.; Kundu, S.C.; Wang, L. Exploring natural silk protein sericin for regenerative medicine: An injectable, photoluminescent, cell-adhesive 3D hydrogel. *Sci. Rep.* **2014**, *4*, 7064. [CrossRef] [PubMed]
34. Liu, J.; Qi, C.; Tao, K.; Zhang, J.; Zhang, J.; Xu, L.; Jiang, X.; Zhang, Y.; Huang, L.; Li, Q. Sericin/dextran injectable hydrogel as an optically trackable drug delivery system for malignant melanoma treatment. *ACS Appl. Mater. Interfaces* **2016**, *8*, 6411–6422. [CrossRef] [PubMed]
35. Jalal, R.; Goharshadi, E.K.; Abareshi, M.; Moosavi, M.; Yousefi, A.; Nancarrow, P. ZnO nanofluids: Green synthesis, characterization, and antibacterial activity. *Mater. Chem. Phys.* **2010**, *121*, 198–201. [CrossRef]
36. Kasemets, K.; Ivask, A.; Dubourguier, H.-C.; Kahru, A. Toxicity of nanoparticles of ZnO, CuO and TiO_2 to yeast Saccharomyces cerevisiae. *Toxicol. In Vitro* **2009**, *23*, 1116–1122. [CrossRef] [PubMed]
37. Wu, J.-H.; Wang, Z.; Xu, S.-Y. Preparation and characterization of sericin powder extracted from silk industry wastewater. *Food Chem.* **2007**, *103*, 1255–1262. [CrossRef]
38. Wang, Y.; Cai, R.; Tao, G.; Wang, P.; Zuo, H.; Zhao, P.; Umar, A.; He, H. A novel Ag NPs/sericin/agar film with enhanced mechanical property and antibacterial capability. *Molecules* **2018**, *23*, 1821. [CrossRef]
39. Guan, J.; Porter, D.; Vollrath, F. Thermally induced changes in dynamic mechanical properties of native silks. *Biomacromolecules* **2013**, *14*, 930–937. [CrossRef]
40. Pal, S.; Tak, Y.K.; Song, J.M. Does the antibacterial activity of silver nanoparticles depend on the shape of the nanoparticle? A study of the gram-negative bacterium Escherichia coli. *Appl. Environ. Microbiol.* **2007**, *73*, 1712–1720. [CrossRef]

Sample Availability: Samples of the compounds are not available from the authors.

molecules

MDPI

Article

A Novel AgNPs/Sericin/Agar Film with Enhanced Mechanical Property and Antibacterial Capability

Yejing Wang [1,2,†], Rui Cai [1,†], Gang Tao [2], Peng Wang [1], Hua Zuo [3], Ping Zhao [2,4], Ahmad Umar [5] and Huawei He [2,4,*]

1. College of Biotechnology, Southwest University, Beibei, Chongqing 400715, China; yjwang@swu.edu.cn (Y.W.); cairui0330@email.swu.edu.cn (R.C.); modelsums@email.swu.edu.cn (P.W.)
2. State Key Laboratory of Silkworm Genome Biology, Southwest University, Beibei, Chongqing 400715, China; taogang@email.swu.edu.cn (G.T.); zhaop@swu.edu.cn (P.Z.)
3. College of Pharmaceutical Sciences, Southwest University, Beibei, Chongqing 400715, China; zuohua@swu.edu.cn
4. Chongqing Engineering and Technology Research Center for Novel Silk Materials, Southwest University, Beibei, Chongqing 400715, China
5. Department of Chemistry, College of Science and Arts and Promising Centre for Sensors and Electronics Devices, Najran University, P.O. Box 1988, Najran 11001, Saudi Arabia; umahmad@nu.edu.sa

* Correspondence: hehuawei@swu.edu.cn; Tel.: +86-23-6825-1575

† These authors have equal contributions to this work.

Received: 17 June 2018; Accepted: 11 July 2018; Published: 23 July 2018

Abstract: Silk sericin is a protein from a silkworm's cocoon. It has good biocompatibility, hydrophilicity, bioactivity, and biodegradability. However, sericin could not be used in biomedical materials directly because of its frangible characteristic. To develop multifunctional sericin-based materials for biomedical purposes, we prepared a sericin/agar (SS/agar) composite film through the blending of sericin and agar and repetitive freeze-thawing. Then, we synthesized silver nanoparticles (AgNPs) in situ on the surface of the composite film to endow it with antibacterial activity. Water contact angle, swelling and losing ratio, and mechanical properties analysis indicated that the composite film had excellent mechanical property, hydrophilicity, hygroscopicity, and stability. Scanning electron microscopy and X-ray photoelectron spectroscopy analysis confirmed the successful modification of AgNPs on the composite film. X-ray powder diffraction showed the face-centered cubic structures of the AgNPs. This AgNPs modified composite film exhibited an excellent antibacterial capability against *Escherichia coli* and *Staphylococcus aureus*. Our study develops a novel AgNPs/sericin/agar composite film with enhanced mechanical performance and an antimicrobial property for potential biomedical applications.

Keywords: silk sericin; agar; silver nanoparticles; antimicrobial activity

1. Introduction

Recently, it appears promising to combine the outstanding properties of natural protein and polysaccharides for biomedical applications [1]. As a natural macromolecular protein, silk sericin (SS) is composed of 25% of the cocoon that is produced by the silkworm [2,3]. Sericin is composed of 18 different amino acids [4]. As sericin has a content of serine that is about 33.43% [5], it is considered as a natural moisturizing factor for the human skin due to its excellent moisture retention capacity [6]. Sericin is not expensive, is readily available, is biocompatible, and biodegradable [7–9]. Sericin is a biologically active substance in wound dressings due to its diverse activities such as ROS clearance, anti-tyrosinase, and its immunomodulatory capacity [10,11]. In addition, sericin can promote skin keratinocytes and fibroblasts adhesion and proliferation [12–15], especially for cell growth and migration [16,17], which makes it favorable for wound dressing and tissue engineering applications [18–21]. Sericin can efficiently promote wound healing by accelerating collagen deposition and the re-epithelialization of skin tissue [22]. However, sericin has a large amount of disordered structures [23,24], resulting in its poor mechanical performance [25]. Thus, crosslinking, blending, or copolymerizing with other substances is often applied to overcome the brittleness of sericin [26].

Agar is a kind of unbranched polysaccharide that is produced by seaweed [27]. It can spontaneously form gels at a very low concentration. Hence, it is widely applied in the food industry, the pharmaceutical industry, and in cosmetics [28,29]. In addition, agar is applied in biomaterials for its high mechanical strength and biocompatibility [30]. It is promising to combine the outstanding properties of sericin and agar to produce a novel composite material with suitable mechanical properties and good biocompatibility for biomedical applications such as tissue engineering and wound healing.

Surface immobilization with an antimicrobial agent is a common method to prepare antibacterial materials. Silver nanoparticle (AgNP) has broad-spectrum antimicrobial activities [31–34] and rarely leads to drug resistance [35]. Also, AgNP has good cytocompatibility [36,37]. AgNP has an anti-inflammatory effect which is beneficial for wound healing [38,39]. AgNPs are commonly modified on materials after synthesis. The traditional procedure is too complicated and time-consuming to be abandoned [40–42]. The biosynthesis of AgNPs is a safe and environmentally friendly method and has received increasing attention [43,44]. However, biosynthesis is too difficult to ensure the in situ synthesis of AgNPs. UV irradiation is usually used to in situ synthesize AgNPs on the surface of materials [45–47]. Here, we blended sericin and agar to prepare a SS/agar film through freeze-thawing for four cycles. Then, the AgNPs were immobilized on the composite film for their antimicrobial purpose.

2. Results and Discussion

Here, we prepared a SS/agar film through blending without an extra cross-linking agent. Sericin was extracted into water through autoclave. A high temperature (60 °C) promoted the blending of sericin and agar. As the temperature decreased, the mixture gelled to form a novel SS/agar film. UV-assisted synthesis is a facile and green approach to synthesize AgNPs in situ on the surface of a material without chemical cross-linking reagents. Here, UV irradiation was used to promote the synthesis of AgNPs. Then, the antimicrobial property of the composite film was evaluated by inhibition zone and bacterial growth assays. The diagram of the fabrication and antimicrobial analysis of AgNPs/SS/ agar film are shown in Figure 1.

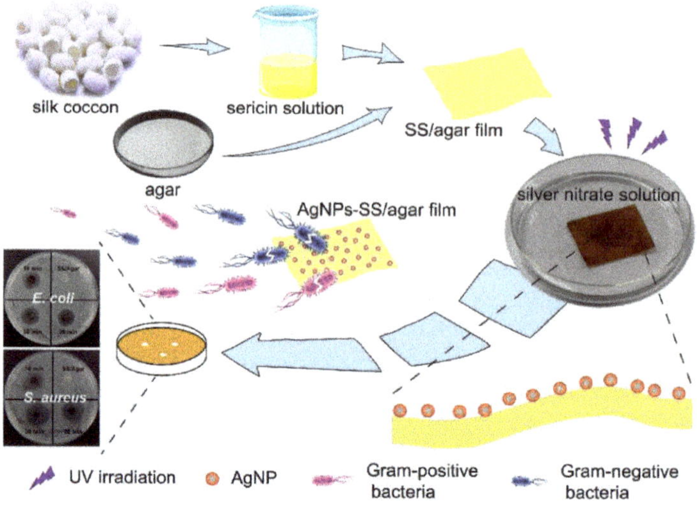

Figure 1. A diagram of the fabrication and antimicrobial test of the AgNPs/SS/agar film.

To determine the effects of the agar contents on the performance of the SS/agar film, the mechanical performance of the SS/agar films with different agar contents were determined. The stress and strain of the SS/agar film increased with increasing agar content in the dry state (Figure 2a,b), respectively. The sericin film failed to test as it was too brittle. The pure agar film had the highest stress and strain. The stress and strain of the SS/agar film (1:1) were 74.3 ± 10.6 MPa and 19.4 ± 4.6%, respectively, which were the highest of the tested films. Figure 2c,d showed a similar tendency in a wet state. The result suggested that agar could significantly improve the rigidity and flexibility of the sericin film. In a wet state, the strain of the SS/agar film (1:1) was 30.3 ± 1.7%, which was higher than that of the film in a dry state, indicating that SS/agar film had better flexibility in the wet state than in the dry state. Our result suggested that the prepared SS/agar film had good rigidity and flexibility, which may have a potential application in biomaterials such as wound dressing and tissue engineering [48–50].

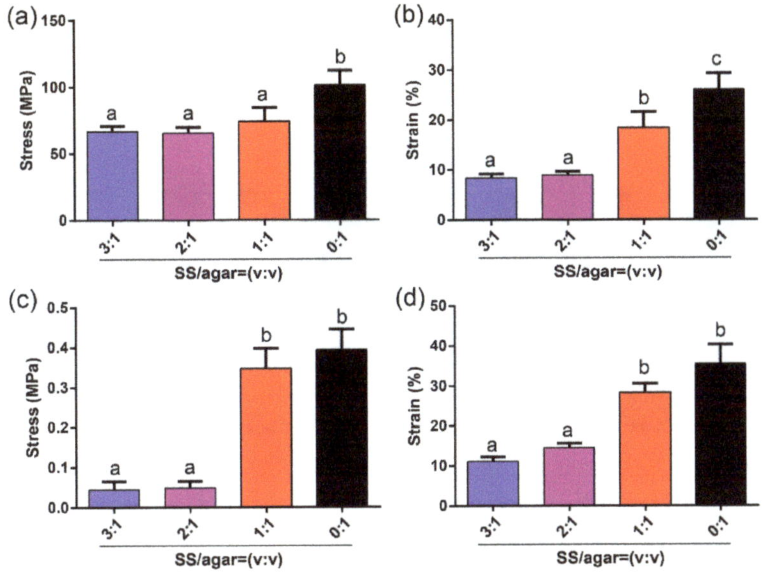

Figure 2. Mechanical properties of the SS/agar films with different ratios (SS: agar = 3:1, 2:1, 1:1, 0:1). The stress of the SS/agar film in a dry state (**a**) and wet state (**c**); The strain of the SS/agar film in a dry state (**b**) and wet state (**d**). For statistical analysis, the bars that are labeled with different lowercase letters (a, b, c) are significantly different, and the same letter indicates no significant difference ($p < 0.05$).

The water contact angle is an important feature of a material's surface hydrophobicity. A water contact angle that is less than 90° indicates that the material is hydrophilic. The agar film had a water contact angle of 84.3°. Increasing sericin's content resulted in a decrease of the water contact angle from 84.3° to 38.8° (Figure 3), indicating that sericin could significantly improve the hydrophilicity of the agar film. This is related to the hydrophilicity of sericin as it contains a great deal of hydrophilic groups.

Swellability indicates the water absorption capacity. The pure agar film had a swelling ratio of 1599.1 ± 84.7% after 12 h (Figure 4a), indicating its good hygroscopicity [27]. The swelling ratio of the SS/agar film increased with the increase of agar content, suggesting that agar could improve the swellability of the SS/agar film. The SS/agar film (1:1) had a swelling ratio of 472.3 ± 9.2% after 12 h (Figure 4a), suggesting that it had good hygroscopicity. The excellent hydrophilicity and swellability are beneficial to the SS/agar film as a wound dressing to keep a moisture microenvironment around the wound interface to promote wound healing.

The losing ratio is an indicator of a material's stability. The pure agar film had the lowest losing ratio in all of the films. The losing ratio of the SS/agar film decreased with the increase of agar content

(Figure 4b), suggesting that agar could improve the stability of the SS/agar film. The SS/agar film (1:1) had a losing ratio of 10.6 ± 0.8% after 24 h (Figure 4b), indicating that it had good stability.

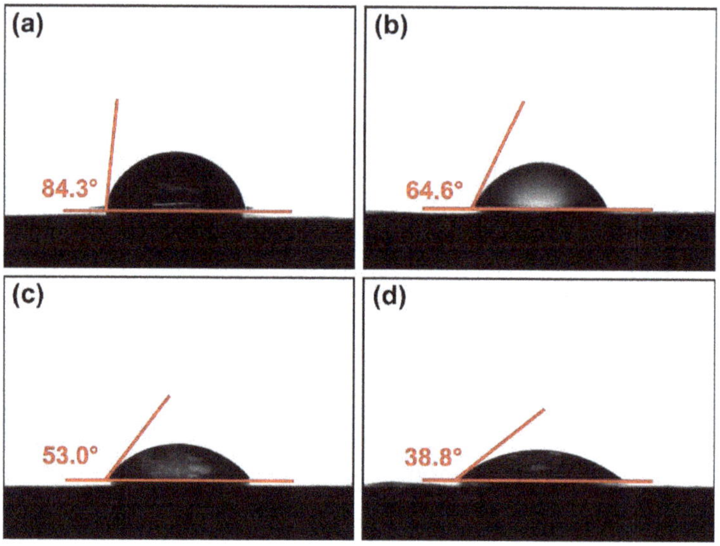

Figure 3. The water contact angle of the SS/agar films. SS: agar = 0:1 (**a**), SS: agar = 1:1 (**b**), SS: agar = 2:1 (**c**), and SS: agar = 3:1 (**d**).

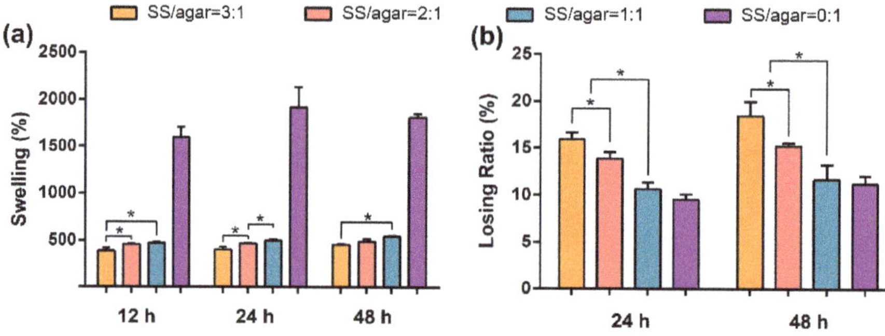

Figure 4. The swelling (**a**) and losing ratio (**b**) of the different SS/agar films. * indicates $p < 0.05$.

Stability is very important for a wound dressing. The mass loss of the SS/agar film in the PBS buffer (pH 7.4) was used to evaluate its stability. After 80 days, the agar film and the SS/agar film (1:1) lost about 48.5% and 55.6% of the initial mass (Figure 5), respectively. The mass loss ratio of the SS/agar film increased with increasing sericin content. Our result indicated the excellent stability of the SS/agar film.

Taken together, the results suggested that the SS/agar film (1:1) had good mechanical performance, excellent hydrophilicity, hygroscopicity, and stability. As a result, the SS/agar film (1:1) was selected for the following experiments.

The surface morphologies of the composite films were characterized by scanning electron microscopy (SEM). As shown in Figure 6a,b, the SS/agar film had a smooth and uniform surface without any defects, indicating that the agar and sericin were well blended. Some small dots appeared on the SS/agar film after UV irradiation for 10 min (Figure 6c,d), indicating the successful synthesis of AgNPs. The synthesized

AgNPs had a size range of 20–50 nm (Figure 6d). Increasing the irradiation time promoted AgNPs synthesis (Figure 6e–h). After UV irradiation for 30 min and 60 min, most of the AgNPs had a size range of 20–60 nm and 20–80 nm, respectively (Figure 6f–h). This result suggested that UV irradiation time could regulate AgNPs synthesis. EDS spectrum showed a featured peak of silver element (Figure 6i), indicating the existence of silver on the AgNPs/SS/agar film. The other featured peaks, such as carbon, nitrogen, and oxygen, were suggested to be from the sericin and agar of the film (Figure 6i).

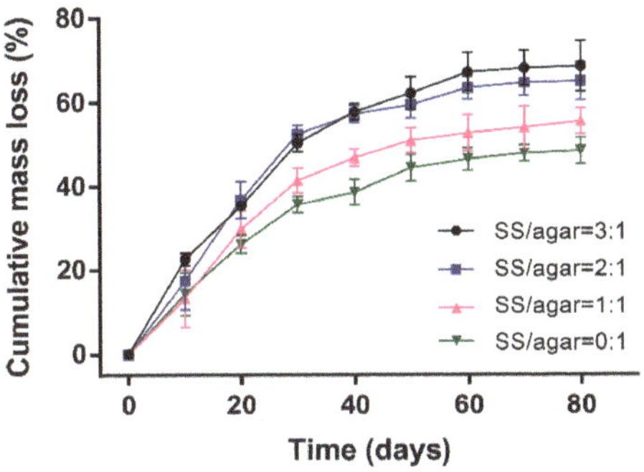

Figure 5. The mass loss of the SS/agar films.

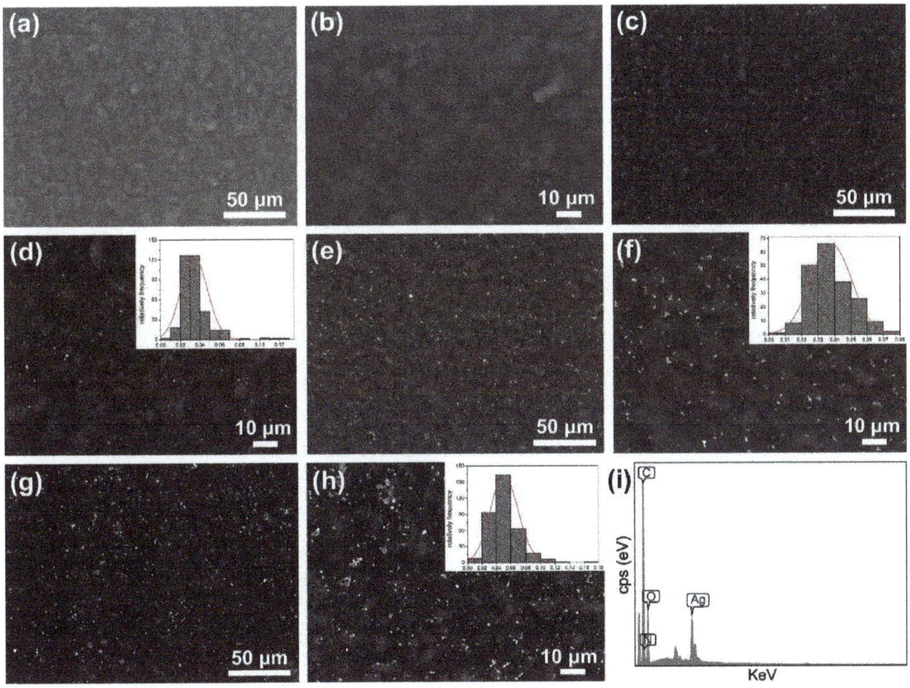

Figure 6. The surface morphology of the different films. The SS/agar film (**a**,**b**) and AgNPs/SS/agar films after UV irradiation for 10 min (**c**,**d**), 30 min (**e**,**f**), and 60 min (**g**,**h**), respectively; inset in (**d**,**f**,**h**): the size distribution of the AgNPs. (**i**) Elemental composition analysis of the AgNPs/SS/agar film by EDS.

X-ray powder diffraction (XRD) was performed to characterize the composite films, as shown Figure 7a. The peak that was located at 19.4° and 13.9° could be assigned to the featured pattern of sericin [51] and the crystalline diffraction of agar [28,30], respectively. The AgNPs/SS/agar film had a significant XRD peak at 38.2°, indicating that the synthesized AgNPs had the crystal planes (111) structure [52,53]. The excellent crystalline structure of the AgNPs implied its efficient antimicrobial activity.

The FT-IR spectra showed the featured peaks of sericin at 1619 cm^{-1} and 1521 cm^{-1} (Figure 7b), which are assigned to the amide I and II of the β-sheet structure [54,55]. The specific bands of agar at 3350 cm^{-1} and 2920 cm^{-1} were associated with the stretching vibration of the O-H and C-H bonds, respectively. The bands at 1640 cm^{-1} and 1373 cm^{-1} were assigned to the stretching vibration of the peptide bonds. The bands at 1072 cm^{-1} and 1045 cm^{-1} were related to the vibration of the C-O bond [28]. After blending with agar, the amide I, II, and III bands of sericin did not change, indicating that the blending did not obviously affect sericin's structure. In addition, after the AgNPs modification on the surface of the SS/agar film, the amide bands of sericin did not change, suggesting that the modification had no effect on the SS/agar film's structure.

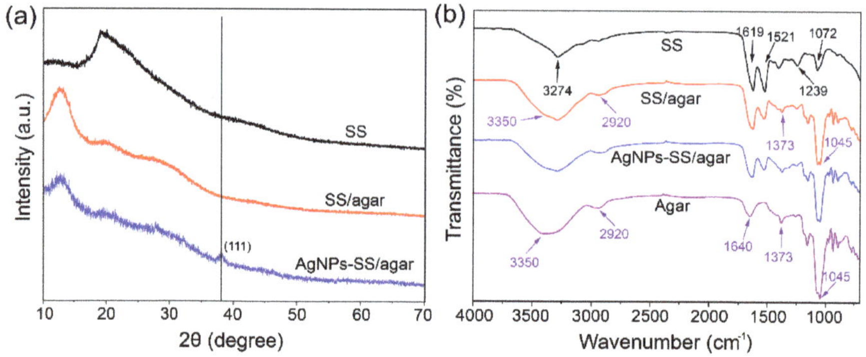

Figure 7. The characterization of different films. (**a**) XRD spectra; (**b**) FT-IR spectra.

X-ray photoelectron spectroscopy (XPS) was carried out to reveal the chemical valence of the synthesized silver. Ag ($3d_{5/2}$) and Ag ($3d_{3/2}$) had the binding energies of 368.08 eV and 374.08 eV (Figure 8), respectively, indicating the formation of Ag^0 [56,57]. Further, the Ag (3d) peak was analyzed by a deconvolution algorithm. Ag, Ag_2O, and AgO have the binding energies of 368.5 eV, 368.3 eV, and 367.7 eV, respectively. Our result suggested that the synthesized AgNPs were composed of about 95% Ag^0, 1% Ag_2O, and 4% AgO.

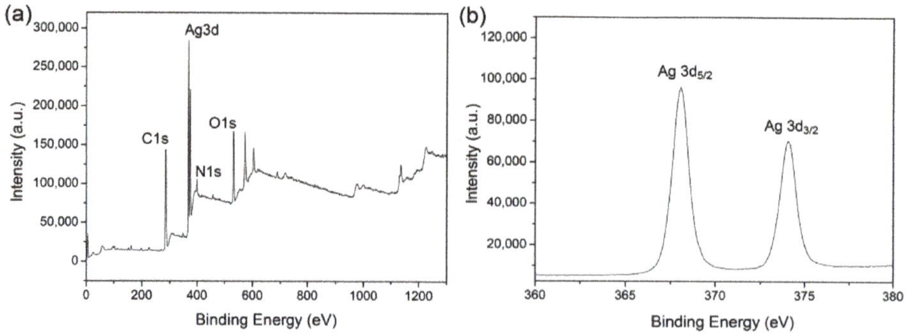

Figure 8. XPS spectra of the AgNPs/SS/agar film (**a**) and the deconvolution of the Ag (3d) peak (**b**).

E. coli and *S. aureus* were chosen as the model of Gram-negative and Gram-positive bacteria to assess the bactericidal activities of the AgNPs/SS/agar film. The inhibition zone assay showed that no inhibition rings appeared in the presence of the SS/agar film, either on *E. coil* or *S. aureus* agar plates. Whereas, the AgNPs/SS/agar films formed obvious inhibition rings on *E. coil* or *S. aureus* agar plates (Figure 9), indicating the excellent antimicrobial activity of the composite film. We measured the diameter of the inhibition rings that were formed, and these are listed in Table 1. There was a statistical difference between the groups for *S. aureus*, but there was no significant difference between the groups for *E. coli*. This result showed that the diameter of the inhibition ring increased with the increase of UV irradiation time, indicating that increasing the UV irradiation time promoted the synthesis of AgNPs and enhanced the antibacterial capability of the AgNPs/SS/agar film on *E. coli* or *S. aureus*.

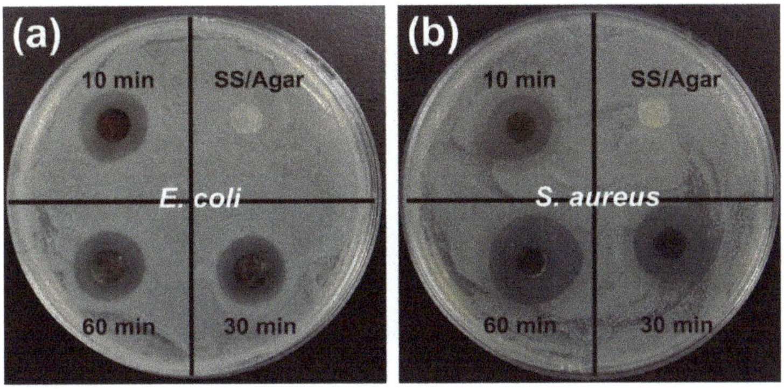

Figure 9. The inhibition zone assay of the SS/agar and AgNPs/SS/agar films. (**a**) *E. coli* and (**b**) *S. aureus*.

Table 1. The diameters of the formed inhibition rings by the AgNPs/SS/agar films with different UV irradiation times.

	SS/Agar (cm)	UV 10 min (cm)	UV 30 min (cm)	UV 60 min (cm)
E. coil	NA	1.67 ± 0.04^{a}	1.69 ± 0.02^{a}	1.74 ± 0.05^{a}
S. aureus	NA	1.70 ± 0.21^{a}	1.81 ± 0.04^{b}	2.21 ± 0.04^{c}

NA, no inhibition ring. The superscript a, b, and c indicate that the data between the groups are statistically different. The superscript with the same letter indicates no statistical difference between the groups. $p < 0.05$.

A bacterial growth assay was performed to further measure the antibacterial activity of the composite film. The bacterial growth showed a similar tendency in the presence of the SS/agar film with that of the control. Nevertheless, in the presence of the AgNPs/SS/agar films, the bacterial growth was obviously inhibited, either for *E. coil* or *S. aureus* (Figure 10). This result was in good agreement with that of the inhibition zone assay, suggesting that the AgNPs/SS/agar film had excellent antimicrobial activity against *E. coil* and *S. aureus*. The AgNPs/SS/agar film with UV irradiation for 60 min showed the best inhibitory effect on the bacterial growth among all of the tested films, either for *E. coli* or *S. aureus*. Given the results of the inhibition zone and bacterial growth assay, the AgNPs modified SS/agar film with UV irradiation for 60 min had the best inhibitory effect on bacteria.

To compare the antibacterial property of Ag (I) ions and AgNPs on *E. coli*, a bacterial colony counting assay was performed. The result showed that the number of bacterial colonies that were treated by the $AgNO_3$/SS/agar film was much more than that of the bacterial colonies that were treated by the AgNPs/SS/agar film (Figure 11), indicating that the AgNPs had a more significant antibacterial effect than the Ag (I) ions on *E. coli*.

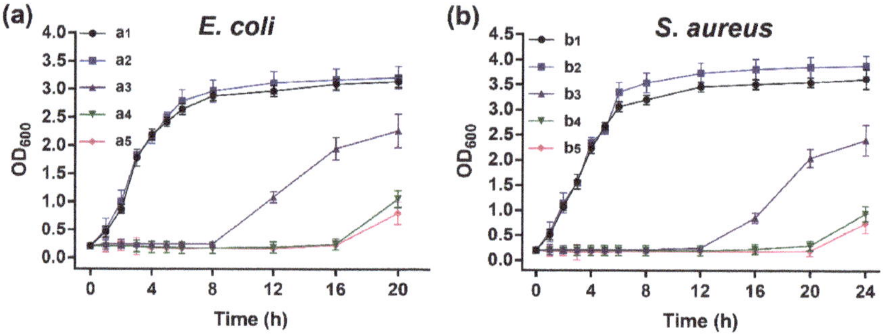

Figure 10. Bacterial growth assay. (**a**) *E. coli* and (**b**) *S. aureus*. a1, b1, control; a2, b2, SS/agar film; a3–b5, AgNPs/SS/agar films with UV irradiation for 10 min (a3, b3), 30 min (a4, b4), and 60 min (a5, b5).

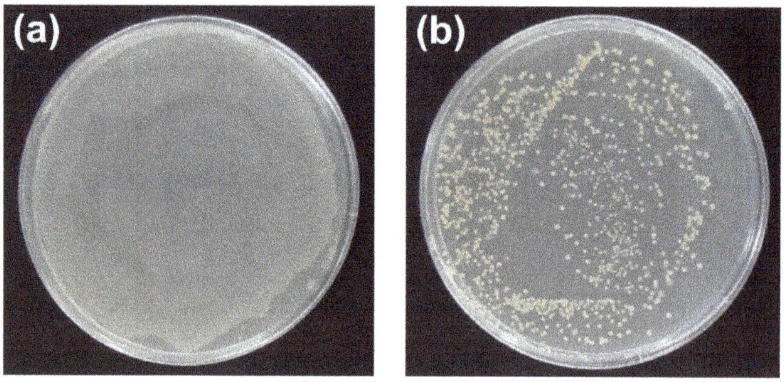

Figure 11. A comparison of the bacterial colony numbers that were treated by the $AgNO_3$/SS/agar film (**a**) and the AgNPs/SS/ agar film (**b**).

3. Materials and Methods

3.1. Materials and Reagents

Silkworm cocoons were kindly provided by the State Key Laboratory of Silkworm Genome Biology, Southwest University, China. In our laboratory, we reared a practical strain of silkworm 872 with fresh mulberry leaves at 25 °C under a photoperiod of 12 h light and 12 h dark and 75% relative humidity. We collected the cocoons for sericin preparation. Agar and $AgNO_3$ (99.99%) were products of Aladdin (Shanghai, China). Ultrapure water that was used in the experiment was prepared by a Milli-Q system (Millipore, Burlington, MA, USA).

3.2. Fabrication of the SS/Agar Film

Sericin was obtained from the silk cocoon as previously reported [58]. The silk cocoons were sliced and autoclaved at 121 °C and 103.4 kPa for 15 min to extract sericin into the water. The insoluble silk fibroin was removed by filtration. The sericin solution was lyophilized to sericin powder. The sericin powder was dissolved in water to become 2% (w/v) sericin solution at 60 °C. The agar was dissolved in water with continuous stirring at 85 °C. Different volume ratios of sericin (2%, w/v) and agar (2%, w/v) were well blended at 60 °C for 15 min and were then air-dried at 37 °C to become the SS/agar film.

3.3. AgNPs Synthesis

AgNPs synthesis with the assistance of UV irradiation is a popular method [59,60]. First, the SS/agar film was soaked in $AgNO_3$ solution (50 mM) at 25 °C, then the film and $AgNO_3$ were exposed to UV irradiation for 10, 30, or 60 min. Ag^+ was reduced to AgNPs in situ on the SS/agar film with the assistance of UV light. Finally, the AgNPs modified SS/agar film was removed from the $AgNO_3$ solution and was air-dried at 37 °C.

3.4. Mechanical Property

The mechanical performance of the SS/agar films were evaluated as the stress and strain that was measured on an AG-Xplus universal machine (Shimadzu, Kyoto, Japan). Each film was sliced into a long strip (5.5 cm × 1 cm). The strip's thickness was determined by an Olympus microscope. Each film was measured at least 10 times and the average value was converted to the stress and strain [61].

3.5. Wettability

The film's wettability was assessed by the water contact angle which was measured on a Krüss DSA100 system (Hamburg, Germany) at 25 °C. To do this, 4 μL of water was dropped on the surface of the film. The complete process of water absorption at five different positions was recorded for each film. The water contact angle was calculated and averaged from five independent measurements.

3.6. Swelling and Losing Ratio

The swelling ratio was determined to measure the film's hygroscopicity, as described by Mandal et al. [62]. After weighing, the dried films (3 cm × 3 cm) were immersed in 10 mL of PBS buffer (pH 7.4) for 12–48 h at 37 °C. Then, the films were taken out from the PBS buffer and were weighed after the removal of the extra water on the surface of the films. The mass of the dried and swollen films were weighed as W_1 and W_2, respectively. Each film was measured at least five times. The swelling ratio was calculated using the Formula (1):

$$S(\%) = \frac{W_2 - W_1}{W_1} \times 100\% \tag{1}$$

The losing ratio was measured to determine the stability of the film. Each sample (3 cm × 3 cm) was immersed in 10 mL of PBS buffer (pH 7.4) for 24–48 h at 37 °C. Then, the film was separated from the PBS buffer and was dried at 65 °C. The initial mass of the SS/agar film and the residual mass of the film after the treatment were weighed as W_3 and W_4, respectively. Each film was measured in triplicate. The losing ratio was determined using the Formula (2):

$$L(\%) = \frac{W_3 - W_4}{W_3} \times 100\% \tag{2}$$

3.7. Mass Loss

The SS/agar films (3 cm × 3 cm) were soaked in 30 mL of PBS buffer (pH 7.4) at 37 °C after drying at 65 °C for 24 h. The initial mass of the film was weighed as W_5. The PBS buffer was substituted daily to keep it fresh. At different intervals, the films were dried at 65 °C for 24 h after separation from the PBS buffer and were then weighed as W_6. Each film was measured three times. The mass loss of each film was determined using the Formula (3):

$$M(\%) = \frac{W_5 - W_6}{W_5} \times 100\% \tag{3}$$

3.8. Materials Characterization

The morphologies of the films were characterized by a JCM-5000 SEM (JEOL, Tokyo, Japan). The particle size (diameter) of about 1000 particles were measured from the SEM images using the Nano Measurer software. Meanwhile, the energy dispersive spectra (EDS) were recorded on an Oxford INCA X-Max 250 (High Wycombe, UK). XRD was measured on a PANalytical X'Pert X-ray diffractometer (Almelo, The Netherlands). 2θ was set from 10° to 80°. The FT-IR spectra were collected on a Thermofisher Nicolet iz10 spectrometer (Framingham, MA, USA). The wavenumber was from 4000 cm^{-1} to 800 cm^{-1} and the resolution was 2 cm^{-1}.

3.9. Inhibition Zone Assay

The assay was carried out as previously described [63]. *E. coli* and *S. aureus* were cultured at 37 °C to have an optical density (OD) at 600 nm (OD_{600}) of 0.2, and were then spread on agar plates to culture with circular sterile SS/agar or AgNPs/SS/agar films (d = 0.7 cm) at 37 °C. After 12 h, the diameters of the formed inhibition zones were measured to assess the antimicrobial activity of the films.

3.10. Growth Curve Assay

The assay was performed as Tao's protocol [64]. Bacteria were cultured to have an initial OD_{600} of 0.2. Then, the circular sterile SS/agar or AgNPs/SS/agar films (d = 0.7 cm) were added to culture with the bacteria at 37 °C. Bacteria (0.2 mL) were collected to measure OD_{600} at different intervals. Each film was measured in triplicate.

3.11. Bacterial Colony Counting Assay

The SS/agar film was soaked in $AgNO_3$ solution (50 mM) at 25 °C and was then exposed to UV irradiation for 30 min to synthesize AgNPs on the composite film. The prepared film was assessed against *E. coli* for its antibacterial activity. The SS/agar film in the $AgNO_3$ solution without UV irradiation was used as a control. Bacteria were cultured to have an initial OD_{600} of 0.2. Then, the circular sterile film (d = 1 cm) was added to culture with *E. coli* at 37 °C. After 6 h, 100 μL of bacterial suspension was collected and was spread on a nutrient agar plate to culture at 37 °C. After 10 h, the bacterial colonies were counted to compare the antibacterial effect of Ag (I) ions and AgNPs on *E. coli*.

3.12. Statistics

For all of the experiments, at least three independent tests were performed, and the data were presented as mean ± SD. The experimental data were analyzed using Student's *t* test and a linear generalized analysis of variance (ANOVA). For the *t* test, * $p < 0.05$; ** $p < 0.01$. For the ANOVA, different lowercase letters indicate a significant difference between the groups, and the same letter means that there is no significant difference between the groups ($p < 0.05$).

4. Conclusions

In this work, we blended sericin with agar to prepare a SS/agar film with an enhanced mechanical performance. Then, we synthesized AgNPs in situ on the surface of the SS/agar film with the assistance of UV irradiation. The AgNPs modified composite film exhibited excellent hydrophilicity, good mechanical properties, and antimicrobial capacity toward *E. coli* and *S. aureus*. This novel sericin-based composite film with its enhanced mechanical performance and antimicrobial capability could potentially be applied in wound dressing and tissue engineering.

Author Contributions: Y.W., R.C., G.T., and H.H. conceived and designed the experiments; R.C., G.T., and P.W. performed the experiments; Y.W., R.C., and G.T. analyzed the data; H.Z. and P.Z. contributed reagents/materials/analysis tools; R.C. and Y.W. wrote the draft; Y.W. and H.H. supervised the research; A.U. and H.H. revised the manuscript.

Funding: This work was funded by National Natural Science Foundation of China (31572465), the Chongqing Research Program of Basic Research and Frontier Technology (cstc2015jcyj BX0035, cstc2015jcyjA00040), the State Key Program of the National Natural Science of China (31530071), Fundamental Research Funds for the Central Universities (XDJK2018B010, XDJK2018C063), the Graduate Research and Innovation Project of Chongqing (CYB17069, CYS18123) and the Open Project Program of Chongqing Engineering and Technology Research Center for Novel Silk Materials (silkgczx2016003).

Conflicts of Interest: The authors declare no conflict of interest. The founding sponsors had no role in the design of the study, in the collection, analyses, or interpretation of the data, in the writing of the manuscript, and in the decision to publish the results.

References

1. Jătariu Cadinoiu, A.N.; Popa, M.; Curteanu, S.; Peptu, C.A. Covalent and ionic co-cross-linking—An original way to prepare chitosan-gelatin hydrogels for biomedical applications. *J. Biomed. Mater. Res. A* **2011**, *98*, 342–350. [CrossRef] [PubMed]
2. Zhang, Y.Q.; Tao, M.L.; Shen, W.D.; Zhou, Y.Z.; Ding, Y.; Ma, Y.; Zhou, W.L. Immobilization of L-asparaginase on the microparticles of the natural silk sericin protein and its characters. *Biomaterials* **2004**, *25*, 3751–3759. [CrossRef] [PubMed]
3. Takasu, Y.; Yamada, H.; Tsubouchi, K. Isolation of three main sericin components from the cocoon of the silkworm, bombyx mori. *Biosci. Biotechnol. Biochem.* **2002**, *66*, 2715–2718. [CrossRef] [PubMed]
4. Padamwar, M.N.; Pawar, A.P. Silk sericin and its applications: A review. *J. Sci. Ind. Res. India* **2004**, *63*, 323–329.
5. Zhang, Y.Q. Applications of natural silk protein sericin in biomaterials. *Biotechnol. Adv.* **2002**, *20*, 91–100. [CrossRef]
6. Zhaorigetu, S.; Yanaka, N.; Sasaki, M.; Watanabe, H.; Kato, N. Inhibitory effects of silk protein, sericin on UVB-induced acute damage and tumor promotion by reducing oxidative stress in the skin of hairless mouse. *J. Photochem. Photobiol. B Biol.* **2003**, *71*, 11–17. [CrossRef]
7. Dash, B.C.; Mandal, B.B.; Kundu, S.C. Silk gland sericin protein membranes: Fabrication and characterization for potential biotechnological applications. *J. Biotechnol.* **2009**, *144*, 321–329. [CrossRef] [PubMed]
8. Mandal, B.B.; Priya, A.S.; Kundu, S.C. Novel silk sericin/gelatin 3-D scaffolds and 2-D films: Fabrication and characterization for potential tissue engineering applications. *Acta Biomater.* **2009**, *5*, 3007–3020. [CrossRef] [PubMed]
9. Siritientong, T.; Srichana, T.; Aramwit, P. The effect of sterilization methods on the physical properties of silk sericin scaffolds. *AAPS PharmSciTech* **2011**, *12*, 771–781. [CrossRef] [PubMed]
10. Chlapanidas, T.; Faragò, S.; Lucconi, G.; Perteghella, S.; Galuzzi, M.; Mantelli, M.; Avanzini, M.A.; Tosca, M.C.; Marazzi, M.; Vigo, D. Sericins exhibit ROS-scavenging, anti-tyrosinase, anti-elastase, and in vitro immunomodulatory activities. *Int. J. Biol. Macromol.* **2013**, *58*, 47–56. [CrossRef] [PubMed]
11. Bari, E.; Arciola, C.; Vigani, B.; Crivelli, B.; Moro, P.; Marrubini, G.; Sorrenti, M.; Catenacci, L.; Bruni, G.; Chlapanidas, T.; et al. In vitro effectiveness of microspheres based on silk sericin and chlorella vulgaris or arthrospira platensis for wound healing applications. *Materials* **2017**, *10*, 983. [CrossRef] [PubMed]
12. Sasaki, M.; Kato, Y.; Yamada, H.; Terada, S. Development of a novel serum-free freezing medium for mammalian cells using the silk protein sericin. *Biotechnol. Appl. Biochem.* **2005**, *42*, 183–188. [PubMed]
13. Tsubouchi, K.; Igarashi, Y.; Takasu, Y.; Yamada, H. Sericin enhances attachment of cultured human skin fibroblasts. *Biosci. Biotechnol. Biochem.* **2005**, *69*, 403–405. [CrossRef] [PubMed]
14. Terada, S.; Nishimura, T.; Sasaki, M.; Yamada, H.; Miki, M. Sericin, a protein derived from silkworms, accelerates the proliferation of several mammalian cell lines including a hybridoma. *Cytotechnology* **2002**, *40*, 3–12. [CrossRef] [PubMed]
15. Zhang, F.; Zhang, Z.B.; Zhu, X.L.; Kang, E.T.; Neoh, K.G. Silk-functionalized titanium surfaces for enhancing osteoblast functions and reducing bacterial adhesion. *Biomaterials* **2008**, *29*, 4751–4759. [CrossRef] [PubMed]
16. Martinez-Mora, C.; Mrowiec, A.; Garcia-Vizcaino, E.M.; Alcaraz, A.; Cenis, J.L.; Nicolas, F.J. Fibroin and sericin from bombyx mori silk stimulate cell migration through upregulation and phosphorylation of c-jun. *PLoS ONE* **2012**, *7*, e42271. [CrossRef] [PubMed]

17. Greaves, N.S.; Ashcroft, K.J.; Baguneid, M.; Bayat, A. Current understanding of molecular and cellular mechanisms in fibroplasia and angiogenesis during acute wound healing. *J. Dermatol. Sci.* **2013**, *72*, 206–217. [CrossRef] [PubMed]
18. Aramwit, P.; Sangcakul, A. The effects of sericin cream on wound healing in rats. *Biosci. Biotechnol. Biochem.* **2007**, *71*, 2473–2477. [CrossRef] [PubMed]
19. Aramwit, P.; Kanokpanont, S.; De-Eknamkul, W.; Srichana, T. Monitoring of inflammatory mediators induced by silk sericin. *J. Biosci. Bioeng.* **2009**, *107*, 556–561. [CrossRef] [PubMed]
20. Crivelli, B.; Perteghella, S.; Bari, E.; Sorrenti, M.; Tripodo, G.; Chlapanidas, T.; Torre, M.L. Silk nanoparticles: From inert supports to bioactive natural carriers for drug delivery. *Soft Matter* **2018**, *14*, 546–557. [CrossRef] [PubMed]
21. Chlapanidas, T.; Perteghella, S.; Leoni, F.; Faragò, S.; Marazzi, M.; Rossi, D.; Martino, E.; Gaggeri, R.; Collina, S. TNF-α blocker effect of naringenin-loaded sericin microparticles that are potentially useful in the treatment of psoriasis. *Int. J. Mol. Sci.* **2014**, *15*, 13624–13636. [CrossRef] [PubMed]
22. Aramwit, P.; Palapinyo, S.; Srichana, T.; Chottanapund, S.; Muangman, P. Silk sericin ameliorates wound healing and its clinical efficacy in burn wounds. *Arch. Dermatol. Res.* **2013**, *305*, 585–594. [CrossRef] [PubMed]
23. Dash, R.; Ghosh, S.K.; Kaplan, D.L.; Kundu, S.C. Purification and biochemical characterization of a 70 kda sericin from tropical tasar silkworm, Antheraea Mylitta. *Comp. Biochem. Phys. B* **2007**, *147*, 129–134. [CrossRef] [PubMed]
24. Oh, H.; Lee, J.Y.; Kim, M.K.; Um, I.C.; Lee, K.H. Refining hot-water extracted silk sericin by ethanol-induced precipitation. *Int. J. Biol. Macromol.* **2011**, *48*, 32–37. [CrossRef] [PubMed]
25. Nayak, S.; Talukdar, S.; Kundu, S.C. Potential of 2D crosslinked sericin membranes with improved biostability for skin tissue engineering. *Cell Tissue Res.* **2012**, *347*, 783–794. [CrossRef] [PubMed]
26. Zhang, X.H.; Tsukada, M.; Morikawa, H.; Aojima, K.; Zhang, G.Y.; Miura, M. Production of silk sericin/silk fibroin blend nanofibers. *Nanoscale Res. Lett.* **2011**, *6*, 510. [CrossRef] [PubMed]
27. Phan, T.D.; Debeaufort, F.; Luu, D.; Voilley, A. Functional properties of edible agar-based and starch-based films for food quality preservation. *J. Agric. Food Chem.* **2005**, *53*, 973–981. [CrossRef] [PubMed]
28. Wu, Y.; Geng, F.Y.; Chang, P.R.; Yu, J.G.; Ma, X.F. Effect of agar on the microstructure and performance of potato starch film. *Carbohydr. Polym.* **2009**, *76*, 299–304. [CrossRef]
29. Li, H.Y.; Yu, X.J.; Jin, Y.; Zhang, W.; Liu, Y.L. Development of an eco-friendly agar extraction technique from the red seaweed gracilaria lemaneiformis. *Bioresour. Technol.* **2008**, *99*, 3301–3305. [CrossRef] [PubMed]
30. Freile-Pelegrin, Y.; Madera-Santana, T.; Robledo, D.; Veleva, L.; Quintana, P.; Azamar, J.A. Degradation of agar films in a humid tropical climate: Thermal, mechanical morphological and structural changes. *Polym. Degrad. Stab.* **2007**, *92*, 244–252. [CrossRef]
31. Lara, H.H.; Garza-Trevino, E.N.; Ixtepan-Turrent, L.; Singh, D.K. Silver nanoparticles are broad-spectrum bactericidal and virucidal compounds. *J. Nanobiotechnol.* **2011**, *9*, 30. [CrossRef] [PubMed]
32. Tao, G.; Wang, Y.J.; Liu, L.N.; Chang, H.P.; Zhao, P.; He, H.W. Preparation and characterization of silver nanoparticles composited on polyelectrolyte film coated sericin gel for enhanced antibacterial application. *Sci. Adv. Mater.* **2016**, *8*, 1547–1552. [CrossRef]
33. Tao, G.; Cai, R.; Wang, Y.J.; Song, K.; Guo, P.C.; Zhao, P.; Zuo, H.; He, H.W. Biosynthesis and characterization of AGNPS-silk/PVA film for potential packaging application. *Materials* **2017**, *10*, 667. [CrossRef] [PubMed]
34. Cai, R.; Tao, G.; He, H.; Song, K.; Zuo, H.; Jiang, W.; Wang, Y. One-step synthesis of silver nanoparticles on polydopamine-coated sericin/polyvinyl alcohol composite films for potential antimicrobial applications. *Molecules* **2017**, *22*, 721. [CrossRef] [PubMed]
35. Dorjnamjin, D.; Ariunaa, M.; Shim, Y.K. Synthesis of silver nanoparticles using hydroxyl functionalized ionic liquids and their antimicrobial activity. *Int. J. Mol. Sci.* **2008**, *9*, 807–819. [CrossRef] [PubMed]
36. Morones, J.R.; Elechiguerra, J.L.; Camacho, A.; Holt, K.; Kouri, J.B.; Ramirez, J.T.; Yacaman, M.J. The bactericidal effect of silver nanoparticles. *Nanotechnology* **2005**, *16*, 2346–2353. [CrossRef] [PubMed]
37. Skladanowski, M.; Golinska, P.; Rudnicka, K.; Dahm, H.; Rai, M. Evaluation of cytotoxicity, immune compatibility and antibacterial activity of biogenic silver nanoparticles. *Med. Microbiol. Immunol.* **2016**, *205*, 603–613. [CrossRef] [PubMed]
38. Elliott, C. The effects of silver dressings on chronic and burns wound healing. *Br. J. Nurs.* **2010**, *19*, S32–S36. [CrossRef] [PubMed]

39. He, H.; Tao, G.; Wang, Y.; Cai, R.; Guo, P.; Chen, L.; Zuo, H.; Zhao, P.; Xia, Q. In situ green synthesis and characterization of sericin-silver nanoparticle composite with effective antibacterial activity and good biocompatibility. *Mater. Sci. Eng. C* **2017**, *80*, 509–516. [CrossRef] [PubMed]
40. Swope, K.L.; Flicklinger, M.C. The use of confocal scanning laser microscopy and other tools to characterize *Escherichia coli* in a high-cell-density synthetic biofilm. *Biotechnol. Bioeng.* **1996**, *52*, 340–356. [CrossRef]
41. Lee, J.M.; Yu, J.E.; Koh, Y.S. Experimental study on the effect of wavelength in the laser cleaning of silver threads. *J. Cult. Herit.* **2003**, *4*, 157s–161s. [CrossRef]
42. Ma, Z.Y.; Jia, X.; Hu, J.M.; Liu, Z.Y.; Wang, H.Y.; Zhou, F. Mussel-inspired thermosensitive polydopamine-graft-poly(*N*-isopropylacrylamide) coating for controlled-release fertilizer. *J. Agric. Food Chem.* **2013**, *61*, 12232–12237. [CrossRef] [PubMed]
43. Vahabi, K.; Mansoori, G.A.; Karimi, S. Biosynthesis of silver nanoparticles by fungus trichoderma reesei (a route for large-scale production of AGNPS). *Insci. J.* **2011**, *1*, 65–79. [CrossRef]
44. Mohammadinejad, R.; Pourseyedi, S.; Baghizadeh, A.; Ranjbar, S.; Mansoori, G. Synthesis of silver nanoparticles using silybum marianum seed extract. *Int. J. Nanosci. Nanotechnol.* **2013**, *9*, 221–226.
45. Fellahi, O.; Das, M.R.; Coffinier, Y.; Szunerits, S.; Hadjersi, T.; Maamache, M.; Boukherroub, R. Silicon nanowire arrays-induced graphene oxide reduction under UV irradiation. *Nanoscale* **2011**, *3*, 4662–4669. [CrossRef] [PubMed]
46. Wang, X.; Gao, W.; Xu, S.; Xu, W. Luminescent fibers: In situ synthesis of silver nanoclusters on silk via ultraviolet light-induced reduction and their antibacterial activity. *Chem. Eng. J.* **2012**, *210*, 585–589. [CrossRef]
47. Cui, X.; Li, C.M.; Bao, H.; Zheng, X.; Lu, Z. In situ fabrication of silver nanoarrays in hyaluronan/PDDA layer-by-layer assembled structure. *J. Colloid Interface Sci.* **2008**, *327*, 459–465. [CrossRef] [PubMed]
48. Bahrami, S.B.; Kordestani, S.S.; Mirzadeh, H.; Mansoori, P. Poly(vinyl alcohol)—Chitosan blends: Preparation, mechanical and physical properties. *Iran. Polym. J.* **2003**, *12*, 139–146.
49. Cervera, M.F.; Heinamaki, J.; Krogars, K.; Jorgensen, A.C.; Karjalainen, M.; Colarte, A.I.; Yliruusi, J. Solid-state and mechanical properties of aqueous chitosan-amylose starch films plasticized with polyols. *AAPS PharmSciTech* **2004**, *5*, 109.
50. Xu, F.; Wen, T.J.; Lu, T.J.; Seffen, K.A. Skin biothermomechanics for medical treatments. *J. Mech. Behav. Biomed.* **2008**, *1*, 172–187. [CrossRef] [PubMed]
51. Tao, G.; Liu, L.N.; Wang, Y.J.; Chang, H.P.; Zhao, P.; Zuo, H.; He, H.W. Characterization of silver nanoparticle in situ synthesis on porous sericin gel for antibacterial application. *J. Nanomater.* **2016**. [CrossRef]
52. Abbasi, A.R.; Morsali, A. Synthesis and properties of silk yarn containing Ag nanoparticles under ultrasound irradiation. *Ultrason. Sonochem.* **2011**, *18*, 282–287. [CrossRef] [PubMed]
53. Feng, J.J.; Zhang, P.P.; Wang, A.J.; Liao, Q.C.; Xi, J.L.; Chen, J.R. One-step synthesis of monodisperse polydopamine-coated silver core-shell nanostructures for enhanced photocatalysis. *New J. Chem.* **2012**, *36*, 148–154. [CrossRef]
54. Zhang, X.M.; Wyeth, P. Using FTIR spectroscopy to detect sericin on historic silk. *Sci. China Chem.* **2010**, *53*, 626–631. [CrossRef]
55. Ling, S.J.; Qi, Z.M.; Knight, D.P.; Shao, Z.Z.; Chen, X. Synchrotron FTIR microspectroscopy of single natural silk fibers. *Biomacromolecules* **2011**, *12*, 3344–3349. [CrossRef] [PubMed]
56. Maity, K.; Panda, D.K.; Lochner, E.; Saha, S. Fluoride-induced reduction of Ag(i) cation leading to formation of silver mirrors and luminescent Ag-nanoparticles. *J. Am. Chem. Soc.* **2015**, *137*, 2812–2815. [CrossRef] [PubMed]
57. Liao, Y.A.; Wang, Y.Q.; Feng, X.X.; Wang, W.C.; Xu, F.J.; Zhang, L.Q. Antibacterial surfaces through dopamine functionalization and silver nanoparticle immobilization. *Mater. Chem. Phys.* **2010**, *121*, 534–540. [CrossRef]
58. Wu, J.H.; Wang, Z.; Xu, S.Y. Preparation and characterization of sericin powder extracted from silk industry wastewater. *Food Chem.* **2007**, *103*, 1255–1262. [CrossRef]
59. Cai, R.; Tao, G.; He, H.; Guo, P.; Yang, M.; Ding, C.; Zuo, H.; Wang, L.; Zhao, P.; Wang, Y. In situ synthesis of silver nanoparticles on the polyelectrolyte-coated sericin/PVA film for enhanced antibacterial application. *Materials* **2017**, *10*, 967. [CrossRef] [PubMed]
60. He, H.; Cai, R.; Wang, Y.; Tao, G.; Guo, P.; Zuo, H.; Chen, L.; Liu, X.; Zhao, P.; Xia, Q. Preparation and characterization of silk sericin/PVA blend film with silver nanoparticles for potential antimicrobial application. *Int. J. Biol. Macromol.* **2017**, *104*, 457–464. [CrossRef] [PubMed]

61. Guan, J.; Porter, D.; Vollrath, F. Thermally induced changes in dynamic mechanical properties of native silks. *Biomacromolecules* **2013**, *14*, 930–937. [CrossRef] [PubMed]
62. Vazquez, B.; Roman, J.S.; Peniche, C.; Cohen, M.E. Polymeric hydrophilic hydrogels with flexible hydrophobic chains. Control of the hydration and interactions with water molecules. *Macromolecules* **1997**, *30*, 8440–8446. [CrossRef]
63. Schillinger, U.; Lucke, F.K. Antibacterial activity of lactobacillus-sake isolated from meat. *Appl. Environ. Microbiol.* **1989**, *55*, 1901–1906. [PubMed]
64. Pal, S.; Tak, Y.K.; Song, J.M. Does the antibacterial activity of silver nanoparticles depend on the shape of the nanoparticle? A study of the gram-negative bacterium *Escherichia coli*. *Appl. Environ. Microbiol.* **2007**, *73*, 1712–1720. [CrossRef] [PubMed]

Sample Availability: Samples of sericin are available from the authors.

molecules

Article

Vacuum Casting and Mechanical Characterization of Nanocomposites from Epoxy and Oxidized Multi-Walled Carbon Nanotubes

Gerald Singer [1], Philipp Siedlaczek [1], Gerhard Sinn [1], Patrick H. Kirner [1], Reinhard Schuller [1], Roman Wan-Wendner [2] and Helga C. Lichtenegger [1,*]

[1] Institute of Physics and Materials Science, University of Natural Resources and Life Sciences Vienna, Peter-Jordan-Straße 82, 1190 Vienna, Austria; gerald.singer@boku.ac.at (G.S.); philipp.siedlaczek@boku.ac.at (P.S.); gerhard.sinn@boku.ac.at (G.S.); patrick.kirner@students.boku.ac.at (P.H.K.); reinhard.schuller@boku.ac.at (R.S.)

[2] Department of Structural Engineering, Ghent University, 9000 Ghent, Belgium; roman.wanwendner@ugent.be

* Correspondence: helga.lichtenegger@boku.ac.at; Tel.: +43-147-654-89211

Academic Editor: Marinella Striccoli
Received: 31 December 2018; Accepted: 29 January 2019; Published: 31 January 2019

Abstract: Sample preparation is an important step when testing the mechanical properties of materials. Especially, when carbon nanotubes (CNT) are added to epoxy resin, the increase in viscosity complicates the casting of testing specimens. We present a vacuum casting approach for different geometries in order to produce specimens from functional nanocomposites that consist of epoxy matrix and oxidized multi-walled carbon nanotubes (MWCNTs). The nanocomposites were characterized with various mechanical tests that showed improved fracture toughness, bending and tensile properties performance by addition of oxidized MWCNTs. Strengthening mechanisms were analyzed by SEM images of fracture surfaces and in-situ imaging by digital image correlation (DIC).

Keywords: vacuum casting; carbon nanotubes (CNTs); nanocomposite; mechanical properties

1. Introduction

Carbon nanotubes (CNTs) can be used as nanoscale fillers in order to improve the mechanical, electrical and thermal properties of polymers. The resulting so-called nanocomposites display significantly enhanced performance for many applications. For instance as electrically conductive and transparent flexible foils prepared from single-walled carbon nanotubes (SWCNTs) in poly(methyl methacrylate) (PMMA) [1], black paints with very high absorption of light [2] or as matrix components of carbon fiber reinforced composites (CFRPs) with increased stiffness and strength [3,4]. Also the dielectric properties and thermal conductivity of nanocomposites are of importance [5] as well as the electromagnetic wave absorption behavior of composites for high-temperature applications [6,7]. Nanocomposites are also of interest for fiber reinforced composites, thermosets and the particular CNT/epoxy materials. It is well known that surface-functionalization of CNTs and processing are highly important to benefit from the excellent properties of CNTs in a composite material [8]. Many different approaches have been developed in order to maximize the effect of reinforcement. For an epoxy matrix, the functionalization of CNTs with amines [9] or epoxy groups [10,11] is widely used, since good compatibility with the matrix can be achieved. Direct binding between reactive groups and the resin or interfacial adhesion is necessary to improve load-transfer within the nanocomposite, since the graphitic surface of CNT is generally inert and only provides weak interactions. It was shown that the effectiveness of the load-transfer mechanism between functionalized CNT and a polymer matrix can be quantified by the shift of the Raman G band [12]. For low filler content it was shown that oxidized CNTs show similar performance in mechanical tests of CNT/epoxy to amine-functionalized CNTs [13]. The advantage of oxidized CNTs is that oxidation can be carried out in a single step, whereas other functionalization treatments require additional reactions that make the process more complex. Usually, further cleaning steps are required in order to remove introduced contamination from the oxidizing agents.

In the present study, oxidized MWCNTs, produced by a "green" process [3], were dispersed in epoxy resin by calendering on a three-roll mill (TRM), which is one of the most successful techniques for the dispersion of CNTs in viscous matrices [14,15]. The resulting CNT/epoxy nanocomposites were mechanically tested in tensile and bending mode. Fracture mechanics of compact tension (CT) specimens were investigated and the fracture surface was studied by SEM.

Adding CNT to epoxies dramatically increases the viscosity and results in increased final porosity, which is an important aspect in the production of testing specimens. The formation of air bubbles during the casting procedure of specimens strongly influences the mechanical test results and has to be avoided. Therefore, in this study a vacuum casting technique was developed to produce high quality tensile specimens.

2. Results and Discussion

2.1. Dispersion

The MWCNTs oxidized according to an oxidation procedure using 30% H_2O_2 aqueous solution [3], were dispersed in epoxy using a three-roll mill (TRM). The resulting dispersions of 0.5 wt% oxidized MWCNTs in epoxy resin were investigated under the optical light microscope. Figure 1a shows the homogenous distribution of oxidized MWCNTs in the epoxy matrix with a maximal agglomerate size of several microns in diameter (Figure 1b).

Calendering is a known technique for the production of dispersions from viscous resins and nanoparticles, e.g., epoxy with CNTs [8]. It was reported that three-roll milling leads to the existence of both individualized CNTs as well as agglomerates [15]. General information about the dispersion quality can be obtained from light microscopy images (Figure 1) and is typically judged by the size of remaining agglomerates (in this case <5 μm and thus judged as suitable for composite production).

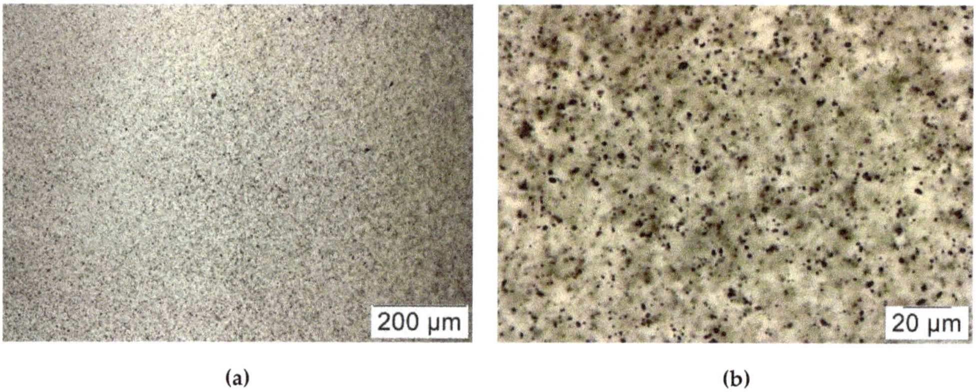

Figure 1. Optical light microscope images of oxidized MWCNT/epoxy dispersions at different magnifications, (**a**) 200 μm; (**b**) 20 μm.

2.2. Vacuum Casting Technique

Tensile specimens were cast in a closed silicone mold according to the illustration in Figure 2. First, funnels were attached to the mold and filled with ~15 g of the mixed components (resin + hardener) each. Then the mold was placed in a vacuum oven and evacuated to ~30 mbar. After the air was fully removed, the oven was carefully ventilated. Due to the atmospheric pressure, the material flows inside the silicone mold. The specimens were cured for 24 h at room temperature and were removed from the mold afterwards.

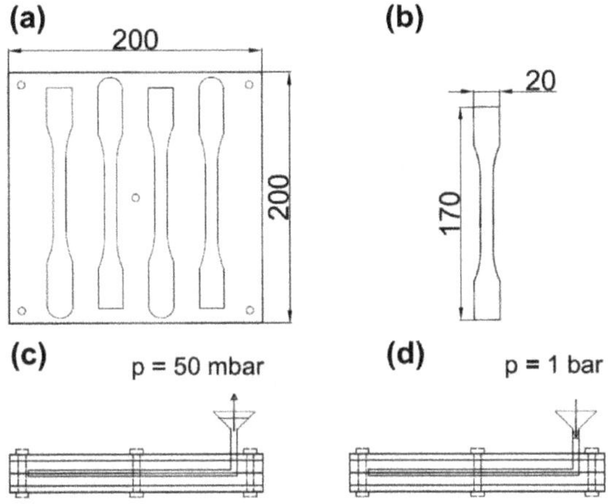

Figure 2. Principle of the vacuum casting technique for tensile specimens: (**a**) Silicone mold for four specimens and (**b**) their dimensions. (**c**) Evacuation of the closed mold with a filled funnel and (**d**) the flow of material inside the mold after ventilation of the vacuum chamber. All measures are given in mm.

Tensile test specimens were much larger compared to the ones used for bending and CT tests and therefore the probability of bubbles within the relevant test area is higher. Furthermore, under tension the whole specimen is under load and will fail at the weakest part, which is given by porosity, whereas for bending and CT tests, only a certain area experiences the maximum load during testing.

2.3. Mechanical Testing

After the resin was mixed with the hardener component, specimens were casted for three different mechanical tests. Tensile specimens were cast using the vacuum casting technique as described in the Materials and Methods section and specimens for bending and CT tests were cast in open silicone molds. The CT specimens required additional drilling and cutting of the notch.

2.3.1. Tensile Tests

Table 1 shows the results of tensile tests of CNT/epoxy nanocomposites (EP/CNT), containing 0.5 wt% oxidized MWCNTs, and neat epoxy (EP) as reference. Adding 0.5 wt% oxidized MWCNTs to the epoxy resin led to an increase of the Young's modulus (E) by +8%, while the tensile strength (σ) was improved by +44%. Also the elongation at break (ε) increased by almost +33% compared to neat epoxy. A small reduction of the Poisson's ratio (ν) by +6% was obtained for CNT/epoxy nanocomposites, which means less contraction in the width of the specimen under axial tensile load. The elastic parameters E and μ were evaluated by digital image correlation (DIC) measurements.

Table 1. Results from tensile tests of neat epoxy (EP) and CNT/epoxy nanocomposites, containing 0.5 wt% oxidized MWCNT (EP/CNT). Mean values are given together with the standard deviation (SD).

	E (MPa)	SD	ν (-)	SD	σ (MPa)	SD	ε (%)	SD
EP	3366	153	0.41	0.02	30.2	2.3	1.12	0.12
EP/CNT	3647	150	0.39	0.05	43.5	4.6	1.49	0.21

Improvement of the Young's modulus by +8% was obtained in this study, which is in the lower range to what was reported in literature, whereas the increases in tensile strength and elongation at break are relatively high [16]. In general, the tensile properties of the CNT/epoxy nanocomposite strongly depend on the amount of filler and their functionalization, the dispersion technique as well as the type of CNT that is used. Particle loadings between 0.1 wt% up to several wt% have been proven to effectively reinforce an epoxy matrix [16]. In a different study relatively high loading of 6 wt% MWCNT were added to epoxy in order to compare the effect on tensile properties of the resulting nanocomposites [17]. The Young's modulus was improved from 3.1 GPa (unreinforced epoxy) to 3.6 GPa and 4.1 GPa by pristine and oxidized MWCNTs, respectively. However, the fracture strength and strain was reduced compared to the reference. At lower CNT loading comparable tensile strength to pure epoxy was reported, while the Young's modulus was improved by +10% (0.1% MWCNTs) and +18% (0.2% MWCNTs).

As a comparison to properly prepared tensile specimens using the presented vacuum technique, mechanical tests of porous specimens that were cast in open silicone molds were performed. The results of specimens with pores were significantly lower and the scattering of the mean values of four tested specimens for Young's modulus and tensile strength was higher (E = 3420 $\pm$ 415 MPa. σ = 27.2 $\pm$ 10.3 MPa, ν = 0.38 $\pm$ 0.03, ε = 0.97 $\pm$ 0.32%) compared to regular specimens. It has to be mentioned that such porous specimens usually have to be excluded from mechanical tests and in this case the samples are used for comparison only. Figure 3 shows the fracture surface of porous tensile specimens. It can be observed that porosity that may exist in the inside of the material (Figure 3a) or at the surface (Figure 3b) initiates the failure.

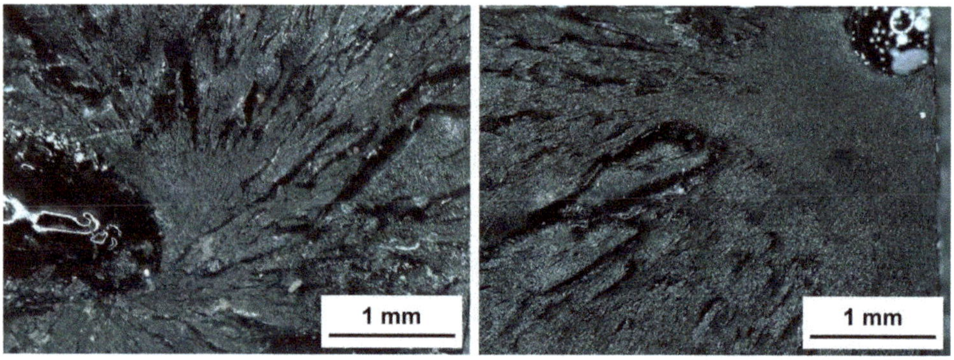

(a) (b)

Figure 3. Optical light microscope images of the fracture surface of porous tensile specimens: Initiation of the crack due to an air bubble (**a**) in the inside of a specimen and (**b**) at the surface of the specimen.

2.3.2. Four-point Bending

The flexural modulus (E_f) was evaluated by DIC measurements and shows an improvement of +12% by adding oxidized MWCNTs to the resin (Table 2). Much higher flexural strength (σ_f) and elongation at break (ε_f) was obtained for nanocomposites with increases of +49% and +84%, respectively.

Table 2. Four-point-bending test results of neat epoxy resin and CNT/epoxy nanocomposites. The numbers represent mean values and the standard deviation (SD).

	E_f (MPa)	SD	σ_f (Mpa)	SD	ε_f (%)	SD
EP	3045	153	66.2	15.5	1.99	0.41
EP/CNT	3401	144	98.5	10.2	3.66	0.89

Similar to the tensile properties, also in bending mode the flexural modulus was improved significantly less compared to the flexural strength and elongation at break. This behavior suggests that the reinforcement mechanism acts more effectively in the plastic deformation than in the elastic range.

2.3.3. Compact Tension (CT) Tests

Results of CT tests are shown in Figure 4. The fracture toughness (K_{Ic}) of investigated nanocomposites was increased by +83% (Figure 4a) and the critical strain energy release rate (G_{Ic}) was improved by +210% (Figure 4b). G_{Ic} was calculated using the following equation:

$$G_{Ic} = \frac{(1-\nu^2)K_{Ic}^2}{E} \quad (1)$$

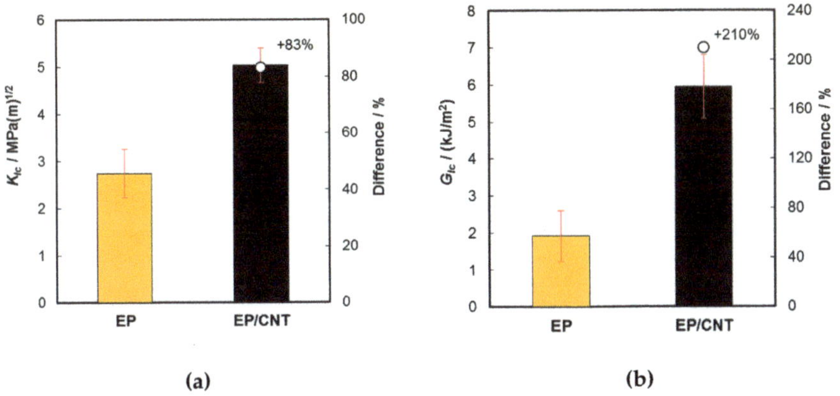

Figure 4. Results obtained from CT tests of neat epoxy (EP) and CNT/epoxy nanocomposites (EP/CNT): (**a**) fracture toughness (K_{Ic}) and (**b**) critical strain energy release rate (G_{Ic}).

Compared to other epoxy resins, the fracture toughness and the critical strain energy release rate of the unmodified and the CNT-modified epoxy in the present study were relatively high. However, the extent of improvement is not unusual when a proper surface functionalization of the CNT is used [16]. Crack bridging, pull-out of CNT and inner tubes as well as rupture of CNT were reported as the main toughening mechanisms in MWCNT/epoxy nanocomposites [18].

2.4. *SEM Investigations*

The micrographs in Figure 5 show fracture surfaces of CT specimens of neat epoxy and CNT/epoxy composite at low (a, b) and high magnification (c, d). At low magnification the crack tip is located at the top of the image. The crack initiation point and propagation direction can be detected by microrivers, radiating from a center. In Figure 5b (CNT/epoxy composite) the crack initiated approximately 500 μm from the crack tip, which indicates a strong bonding at the predetermined breaking point. This appearance is not visible in the sample without CNT (Figure 5a), where the crack initiated directly at the crack tip.

The most obvious difference between neat epoxy and CNT/epoxy composite is the surface roughness around the crack initiation point. It seems that CNT-filled epoxy composite (Figure 5b) had undergone high plastic deformation, which is usually not common for brittle thermoset polymers. From [15] it is known that there are two dominant mechanisms that describe fracture deformation behavior in polymers, which are shear yielding and crazing. Shear yielding is defined by a molecular shear motion at constant volume which leads to plastic deformation, whereas crazing are microcracks originating from microvoids or inhomogeneities. While shear yielding is significant in CNT-dispersed epoxy, sparse features are observed in neat epoxy, indicating a brittle fracture mechanism. Crazing is usually found in brittle thermoplastics and limits plastic deformation. However, multiple crazing leads to toughening and even ductile like behavior can be clearly observed in the CNT/epoxy composite (Figure 5b) as microcracks perpendicular to the propagation direction. Crazes could have arisen from CNT agglomerates, acting as inhomogenities that act as initiators for fine cracks.

At higher magnification (Figure 5d) modified CNT bundles are identified as well dispersed individual units or agglomerated bundles. Compared to neat epoxy (Figure 5c), high roughness and increased plastic deformation can be observed. According to Cha et al. [11], CNT can bridge cracks and increase the fracture toughness by CNT pulling-out. Those mechanisms can be utilized best, if CNT are oriented perpendicular to the crack growth, however randomly dispersed CNT could have a positive effect on fracture toughness already [16]. Rupture and pull-out effects can be exploited in a particularly effective way, if a strong interface between matrix and CNT is provided, e.g., by oxidation of CNT. Since almost all CNTs, protruding out of the resin, appear to have similar length, it can be suggested that CNT pull-out and subsequent rupture was the main mechanism to absorb fracture energy.

The described failure mechanisms were observed for all specimens, suggesting the typical effect of CNT-reinforcement in this case is based on effective load-transfer between oxidized MWCNT and epoxy matrix that results in increased fracture energy due to: (I) high load-bearing capacity of MWCNT, (II) energy-consuming pull-out of CNT and (III) crack branching, leading to higher roughness of the fracture surface and additional plastic deformation.

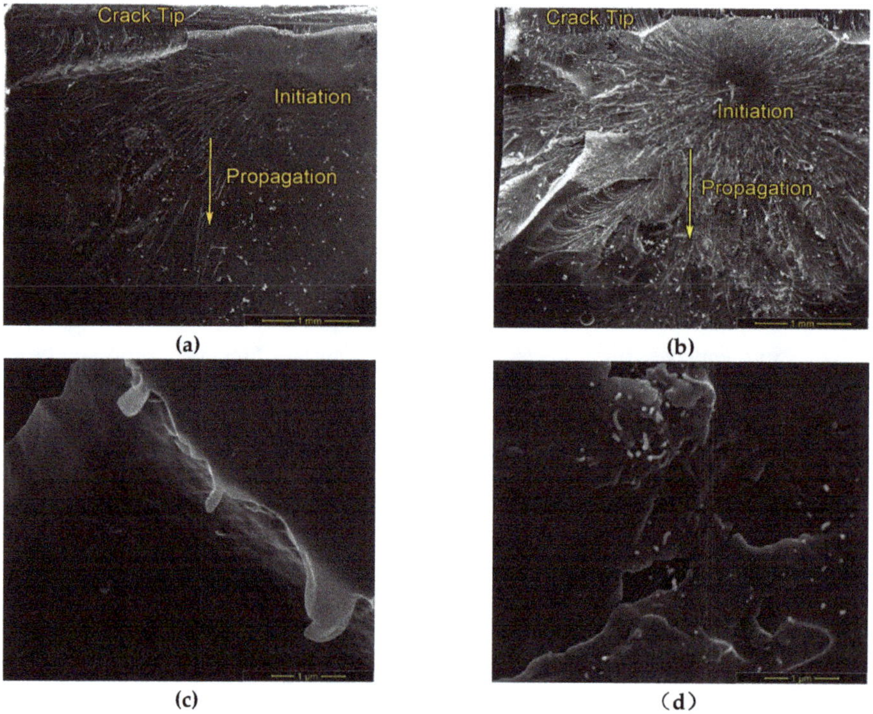

Figure 5. Micrographs from neat epoxy and CNT/epoxy. (**a**) Neat epoxy: Crack initiation started at the edge of the crack tip. (**b**) CNT/epoxy: Crack initiation started 0,5 mm away from the edge. (**c**) Neat epoxy: higher magnification reveals smooth surface. (**d**) CNT/epoxy: higher magnification shows plastic deformation at carbon nanotube agglomerates.

3. Materials and Methods

3.1. Nanocomposites

The epoxy resin was based on bisphenol-A/F epichlorohydrin with *m*-Phenylenebis-(methylamine) hardener. The mixing weight ratio of resin and hardener was 3:1. Pristine MWCNTs (NC7000, Nanocyl, Sambreville, Belgium) with average diameter of ~9.5 nm and 1.5 µm length were oxidized using 30% H_2O_2 aqueous solution according to a previously developed procedure [3]. A pre-dispersion of 0.5 wt% oxidized MWCNT in epoxy resin was produced with a mechanical stirrer. In the next step the fine dispersion was done on a three-roll mill (type 80E, Exakt, Norderstedt, Germany) using the parameters given in Table 3.

Table 3. Dispersion parameters on the three-roll mill.

Step	Gap 1	Gap 2
1	120 µm	40 µm
2	30 µm	10 µm
3	15 µm	5 µm
4	5 µm	10 N/mm

3.2. Mechanical Testing

For tensile tests the vacuum cast specimens according to DIN EN ISO 527 were ground to 4 mm and sprayed with a speckle pattern using white undercoating and black dot pattern. A spindle-driven testing frame (Zwick/Roell, 10 kN, Ulm, Germany) was used at a constant testing speed of 5 mm/min. Four tensile specimens were tested of each type. The surface deformation of the specimen was recorded with a digital image correlation (DIC) system (Q-400, Dantec Dynamics, Skovlunde, Denmark), consisting of two cameras. The images were processed using Istra 4D V4.4.4.694 to calculate the surface mean principal strains.

Specimens for four-point bending tests were casted in an open silicone mold. The dimensions of the bending specimens were 80 × 15 (mm^2) and the thickness was ground to 4 mm. The distance of the support span of the fixture was 66 mm and 22 mm for the loading span. A least four specimens were tested of both, neat and CNT-reinforced epoxy. As already described for the tensile specimens a speckle pattern was created on the side of the specimen in order to record the deformation with DIC. Loading speed was set constant to 2 mm/min.

The compact tension (CT) specimens were casted in an open silicone mold, ground to 4 mm thickness, drilled and notched according to the dimensions in Figure 6. In order to produce precise holes, an in-house built metal template was used in which the specimen was fixed. The notch was cut on a circular saw and before testing, a fine crack was initiated with a fresh blade. The strain was measured with a clip-gage that was attached directly to the specimen on the side of the notch. Force was applied with a constant speed of 0.5 N/mm. For both testing series at least five CT specimens were tested.

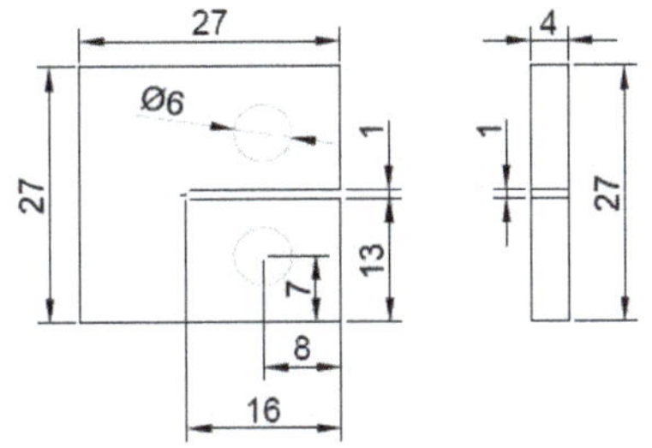

Figure 6. Dimensions (in mm) of the compact tension (CT) specimen.

3.3. Scanning Electron Microscopy (SEM)

For the SEM investigations a 250 FEG (FEI Quanta, Thermo Fisher Scientific, Hillsboro, OR, USA) was used under high vacuum and secondary electron (SE) mode. The fracture surface was sputtered with a thin layer of gold in order to provide sufficient electrical conductivity. Accelerating voltages between 20–30 kV were used for imaging.

4. Conclusions

In this study, a vacuum casting technique for tensile specimens of MWNCT/epoxy nanocomposites was presented as a novel sample preparation method, which is a critical step and usually very challenging due to the high viscosity of the resin. Results of tensile tests were compared to a regular casting technique where porous specimens and reduced properties were observed, which highlights the importance of proper sample preparation. It was shown that adding oxidized MWCNTs to epoxy resin may significantly improve the mechanical properties of the resulting MWCNT/epoxy nanocomposite. In this case, the elastic properties were slightly improved, while the strength and elongation at break were increased in a much more pronounced way. Obviously, the reinforcing mechanism is more important in the plastic deformation than in the elastic region. The fracture toughness of MWCNT-reinforced epoxy and the critical strain energy release rate were enhanced substantially. The fracture surface of CT specimens was investigated with SEM, revealing higher plastic deformations and a rougher surface morphology of MWCNT/epoxy nanocomposite compared to neat epoxy. Furthermore, probably due to the suitable surface modification by oxidation of the CNT, a substantial amount of crack bridging and fiber pull-out was observed, resulting in increased fracture energy and thus improved strength and fracture toughness of MWCNT/epoxy nanocomposites.

Author Contributions: Conceptualization: G.S. (Gerald Singer), H.C.L.; Methodology: G.S. (Gerald Singer), R.S., P.S.; Validation: G.S. (Gerald Singer), P.S.; Formal analysis: G.S. (Gerald Singer), G.S. (Gerhard Sinn); Investigation: G.S. (Gerald Singer), P.S., G.S. (Gerhard Sinn), P.H.K.; Data curation: G.S. (Gerald Singer), G.S. (Gerhard Sinn); writing—original draft preparation: G.S. (Gerald Singer), P.S., P.H.K.; writing—review and editing: H.C.L., R.W.W., R.S.; Supervision: H.C.L., R.W.W., G.S. (Gerhard Sinn); Project administration: H.C.L., R.W.W.

Acknowledgments: The authors would like to thank our technician Daniel Irrasch for his great assistance. Supported by BOKU Vienna Open Access Publishing Fund.

Conflicts of Interest: The authors declare no conflict of interest. The funders had no role in the design of the study; in the collection, analyses, or interpretation of data; in the writing of the manuscript, or in the decision to publish the results.

References

1. Clayton, L.M.; Sikder, A.K.; Kumar, A.; Cinke, M.; Meyyappan, M.; Gerasimov, T.G.; Harmon, J.P. Transparent Poly(methyl methacrylate)/Single-Walled Carbon Nanotube (PMMA/SWNT) Composite Films with Increased Dielectric Constants. *Adv. Funct. Mater.* **2005**, *15*, 101–106. [CrossRef]
2. Lehman, J.; Yung, C.; Tomlin, N.; Conklin, D.; Stephens, M. Carbon nanotube-based black coatings. *Appl. Phys. Rev.* **2018**, *5*, 011103. [CrossRef]
3. Singer, G.; Siedlaczek, P.; Sinn, G.; Rennhofer, H.; Mičušík, M.; Omastová, M.; Unterlass, M.; Wendrinsky, J.; Milotti, V.; Fedi, F.; et al. Acid Free Oxidation and Simple Dispersion Method of MWCNT for High-Performance CFRP. *Nanomaterials* **2018**, *8*, 912. [CrossRef] [PubMed]
4. Singer, G.; Rennhofer, H.; Sinn, G.; Unterlass, M.M.; Wendrinsky, J.; Windberger, U.; Lichtenegger, H.C. Processing of Carbon Nanotubes and Carbon Nanofibers towards High Performance Carbon Fiber Reinforced Polymers. *Key Eng. Mater.* **2017**, *742*, 31–37. [CrossRef]
5. Wang, Z.; Cheng, Y.; Yang, M.; Huang, J.; Cao, D.; Chen, S.; Xie, Q.; Lou, W.; Wu, H. Dielectric properties and thermal conductivity of epoxy composites using core/shell structured Si/SiO_2/Polydopamine. *Compos. Part B: Eng.* **2018**, *140*, 83–90. [CrossRef]
6. Jia, Z.; Lin, K.; Wu, G.; Xing, H.; Wu, H. Recent Progresses of High-Temperature Microwave-Absorbing Materials. *Nano* **2018**, *13*, 1830005. [CrossRef]

7. Lan, D.; Qin, M.; Yang, R.; Chen, S.; Wu, H.; Fan, Y.; Fu, Q.; Zhang, F. Facile synthesis of hierarchical chrysanthemum-like copper cobaltate-copper oxide composites for enhanced microwave absorption performance. *J. Colloid Interface Sci.* **2019**, *533*, 481–491. [CrossRef] [PubMed]
8. Ma, P.-C.; Siddiqui, N.A.; Marom, G.; Kim, J.-K. Dispersion and functionalization of carbon nanotubes for polymer-based nanocomposites: A review. *Compos. Part A: Appl. S.* **2010**, *41*, 1345–1367. [CrossRef]
9. Spitalsky, Z.; Tasis, D.; Papagelis, K.; Galiotis, C. Carbon nanotube–polymer composites: Chemistry, processing, mechanical and electrical properties. *Prog. Polym. Sci.* **2010**, *35*, 357–401. [CrossRef]
10. Sydlik, S.A.; Lee, J.-H.; Walish, J.J.; Thomas, E.L.; Swager, T.M. Epoxy functionalized multi-walled carbon nanotubes for improved adhesives. *Carbon* **2013**, *59*, 109–120. [CrossRef]
11. Martinez-Rubi, Y.; Ashrafi, B.; Guan, J.; Kingston, C.; Johnston, A.; Simard, B.; Mirjalili, V.; Hubert, P.; Deng, L.; Young, R.J. Toughening of epoxy matrices with reduced single-walled carbon nanotubes. *ACS Appl. Mater. Inter.* **2011**, *3*, 2309–2317. [CrossRef] [PubMed]
12. Wang, S.; Liang, R.; Wang, B.; Zhang, C. Load-transfer in functionalized carbon nanotubes/polymer composites. *Chem. Phys. Lett.* **2008**, *457*, 371–375. [CrossRef]
13. Singh, B.P.; Singh, D.; Mathur, R.B.; Dhami, T.L. Influence of Surface Modified MWCNTs on the Mechanical, Electrical and Thermal Properties of Polyimide Nanocomposites. *Nanoscale Res. Lett.* **2008**, *3*, 444–453. [CrossRef]
14. Thostenson, E.T.; Chou, T.-W. Processing-structure-multi-functional property relationship in carbon nanotube/epoxy composites. *Carbon* **2006**, *44*, 3022–3029. [CrossRef]
15. Gojny, F.H.; Wichmann, M.H.G.; Köpke, U.; Fiedler, B.; Schulte, K. Carbon nanotube-reinforced epoxy-composites: Enhanced stiffness and fracture toughness at low nanotube content. *Compos. Sci. Technol.* **2004**, *64*, 2363–2371. [CrossRef]
16. Domun, N.; Hadavinia, H.; Zhang, T.; Sainsbury, T.; Liaghat, G.H.; Vahid, S. Improving the fracture toughness and the strength of epoxy using nanomaterials – a review of the current status. *Nanoscale* **2015**, *7*, 10294–10329. [CrossRef] [PubMed]
17. Delpeux, S.; Benoit, R.; Salvetat, J.P.; Sinturel, C.; Beguin, F.; Bonnamy, S.; Desarmot, G.; Boufendi, L. Functionalization of multiwall carbon nanotubes: Properties of nanotubes-epoxy composites AU–Breton, Y. *Mol. Cryst. Liq. Cryst.* **2002**, *387*, 135–140. [CrossRef]
18. Cha, J.; Jun, G.H.; Park, J.K.; Kim, J.C.; Ryu, H.J.; Hong, S.H. Improvement of modulus, strength and fracture toughness of CNT/Epoxy nanocomposites through the functionalization of carbon nanotubes. *Compos. Part B: Eng.* **2017**, *129*, 169–179. [CrossRef]

Sample Availability: Not available.

molecules

MDPI

Article

Effect of $CaCO_3$ Nanoparticles on the Mechanical and Photo-Degradation Properties of LDPE

Paula A. Zapata [1,*], Humberto Palza [2], Boris Díaz [1], Andrea Armijo [1], Francesca Sepúlveda [1], J. Andrés Ortiz [1], Maria Paz Ramírez [1] and Claudio Oyarzún [1]

1 Grupo Polímeros, Facultad de Química y Biología, Departamento de Ciencias del Ambiente, Universidad de Santiago de Chile, USACH, Santiago 8320000, Chile; borisdiazn@gmail.com (B.D.); andrea.armijot@gmail.com (A.A.); francesca.sepulveda@usach.cl (F.S.); jonathan.ortizn@usach.cl (J.A.O.); maria.ramirezn@usach.cl (M.P.R.); claudio.oyarzun@usach.cl (C.O.)

2 Departamento de Ingeniería Química y Biotecnología, Facultad de Ciencias Físicas y Matemáticas, Universidad de Chile, Beauchef 851, Santiago 8320198, Chile; hpalza@ing.uchile.cl

* Correspondence: paula.zapata@usach.cl

Received: 14 November 2018; Accepted: 21 December 2018; Published: 31 December 2018

check for updates

Abstract: $CaCO_3$ nanoparticles of around 60 nm were obtained by a co-precipitation method and used as filler to prepare low-density polyethylene (LDPE) composites by melt blending. The nanoparticles were also organically modified with oleic acid (O-$CaCO_3$) in order to improve their interaction with the LDPE matrix. By adding 3 and 5 wt% of nanofillers, the mechanical properties under tensile conditions of the polymer matrix improved around 29%. The pure LDPE sample and the nanocomposites with 5 wt% $CaCO_3$ were photoaged by ultraviolet (UV) irradiation during 35 days and the carbonyl index (CI), degree of crystallinity (χ_c), and Young's modulus were measured at different times. After photoaging, the LDPE/$CaCO_3$ nanocomposites increased the percent crystallinity (χ_c), the CI, and Young's modulus as compared to the pure polymer. Moreover, the viscosity of the photoaged nanocomposite was lower than that of photoaged pure LDPE, while scanning electron microscopy (SEM) analysis showed that after photoaging the nanocomposites presented cavities around the nanoparticles. These difference showed that the presence of $CaCO_3$ nanoparticles accelerate the photo-degradation of the polymer matrix. Our results show that the addition of $CaCO_3$ nanoparticles into an LDPE polymer matrix allows future developments of more sustainable polyethylene materials that could be applied as films in agriculture. These LDPE-$CaCO_3$ nanocomposites open the opportunity to improve the low degradation of the LDPE without sacrificing the polymer's behavior, allowing future development of novel eco-friendly polymers.

Keywords: $CaCO_3$ nanoparticles; polyethylene nanocomposites; photoaged polyethylene

---

1. Introduction

Inorganic fillers are incorporated into a polyolefin to form composites with enhanced mechanical, thermal, and barrier properties compared to the polymer matrix [1]. Nanocomposites are a class of filled polymers in which nanometric inorganic fillers are incorporated into the polymer matrix with property enhancements at much lower concentrations than those of microfillers. Calcium carbonate is one of the most commonly used inorganic fillers in thermoplastic polymers, such as poly(vinyl chloride) and polypropylene, to improve their mechanical properties. $CaCO_3$ nanoparticles, in particular, have been incorporated into a polyethylene (PE) matrix by the melting process, increasing Young's modulus with the filler concentration and decreasing both the upper yield point and elongation at break compared to pure PE [2–5]. Although $CaCO_3$ is a well-known filler in polymer composites for mechanical reinforcement, at the nanometric scale it can add other functionalities to the polymer matrix, such as barrier properties and antimicrobial behavior [6,7].

Recently, the effect of the incorporation of nano-particulate calcium carbonate hollow spheres (3, 10 and 25 wt%) in high-density polyethylene (HDPE) by extrusion was studied. They found a crystallinity decrease with increasing filler content. There was found a typical increase of Young's modulus (E) (ca. 17%) with increasing concentration of hollow spheres of $CaCO_3$ filler due to the rigidity of the filler particles and the strong interaction of the filler with the polymer matrix, and it was companied by the corresponding decrease of the upper yield point and elongation at break [2].

One of the major drawbacks of polymer nanocomposites is the high agglomeration of the fillers, which can be reduced by surface modifications. For instance, Lazzeri et al. [8] studied the influence of the organic surface modification in $CaCO_3$ nanoparticles (70 nm) by stearic acid (SA) treatment on the mechanical properties of HDPE composites. Incorporation of 10 vol% of $CaCO_3$ to HDPE increased a rise in yield stress in all composites, but the yield stress decreases with increasing SA content. The author explained this behavior by stating that the addition of SA to the surface of the particles should reduce the stress transfer ability of the interface and even its thickness, leading to a softer interface. The influence of nano-$CaCO_3$ and its surface modification have also been studied in polypropylene (PP) composites prepared by the melting process. In particular, nano-$CaCO_3$ (diameter ca. 44 nm) was modified with stearic acid [1,9,10] and palmitic acid [11], with the addition of modified $CaCO_3$ increasing tensile strength, Young's modulus, and melting point. Another route to improve the dispersion of nano-$CaCO_3$ in PP matrices is by the addition of a small amount of a non-ionic modifier during melt extrusion. In this case Young's modulus increased slightly with amount of $CaCO_3$ load, while the yield strength of PP decreased [12].

On the other hand, to improve the physicochemical properties of the polymer some researchers have treated the surface polymer using thin-layer technology, including oxygen and nitrogen plasma discharge, deposition of functional coatings (i.e., diamond-like carbon (DLC)) among others. For example, low-density polyethylene (LDPE) increased its surface hardness 7 times after layer deposition by DLC coating. Those techniques allowed giving desirable surface properties to the polymer [13].

Despite the relevance for society of the degradation properties of plastic materials, the environmental stability of polymer/calcium carbonate nanocomposites has been barely reported. The effect of nano-$CaCO_3$ on the natural photo-aging degradation of PP was studied outdoors during 88 days [14]. The degradation polymer was studied by Fourier transform infrared (FTIR) spectroscopy and pyrolysis gas chromatography-mass spectroscopy (PGC-eMS). The PP/$CaCO_3$ nanocomposites showed higher photo-degradation than neat PP. The authors explained this behavior as due to the functional groups on the surface of nanoparticles catalyzing the photo-oxidation reaction of PP. There are adsorbed hydroxyl groups on the surface, which is active in photo-chemical reactions. Morreale et al. [15] studied the accelerated weathering behavior of PP/$CaCO_3$ micro- and nanocomposites, showing that the nanosized filler may lead to a faster photo-oxidation rate than that of pure polypropylene. In particular, nanosized calcium carbonate caused faster photodegradation rates than microsized calcium carbonate.

Achieving the bidodegradation of commercial commodity plastics is an enormous environmental challenge due to the increased social demand for higher sustainability processes. The addition of additive/filler to accelerate the photodegradation of these polymers can be associated with an early decrease in the mechanical property even during use. Therefore, nanoparticles able to accelerate the photodegradation together with improving the mechanical behavior can compensate for the latter issue.

Considering what was mentioned above, the present work studies the effect of adding pure and organic-modified $CaCO_3$ nanoparticles into a non-polar LDPE matrix. The effect of different amounts of pure $CaCO_3$ nanoparticles and oleic acid-modified-$CaCO_3$ on the thermal and mechanical properties of polyethylene were studied. The effect of $CaCO_3$ nanoparticles on the photoaging process of LDPE was further investigated.

2. Experiment

2.1. Materials

Polyethylene was purchased from Aldrich, density: 0.925 g cm^{-3}, melt index: 25 g/10 min (190 °C/2.16 kg); impact strength: 45.4 J/m (Izod, ASTM D 256, −50 °C). $CaCO_3$ nanoparticles were synthesized by a precipitation method [16]. Briefly, the reagents used were sodium carbonate, Na_2CO_3 (Merck, Darmstadt, Germany, 99.9%), calcium nitrate, $Ca(NO_3)_2$ (Aldrich, Darmstadt, Germany, 99%), sodium nitrate $NaNO_3$ (Aldrich, 99%), sodium hydroxide, NaOH pellets (Mallinckrodt Chemicals., Dublin, Ireland, ≥98%), and distilled water. Oleic acid (Aldrich, reagent grade, 98%) was used for the modification of the $CaCO_3$ nanoparticles.

2.2. $CaCO_3$ Nanoparticle Synthesis

The $CaCO_3$ nanoparticles were obtained by a method reported by Babou-Kammoe et al. [16]. First, sodium carbonate (Na_2CO_3) (0.042 g) was dissolved in deionized water (80 mL) with sodium hydroxide (NaOH) (1.25 g) and sodium nitrate $NaNO_3$ (0.612 g). In a second step, calcium nitrate ($Ca(NO_3)_2$) (0.944 g) was dissolved in deionized water (80 mL) and the resultant mixture formed a precipitate. The calcium nitrate solution was added dropwise to the sodium carbonate solution with continuous stirring during 4 h at 25 °C. The resultant mixture formed a precipitate which was separated from the water by filtering off. The nanoparticles were dried at 60 °C during 24 h and were characterized.

2.3. Organic Modification of $CaCO_3$ Nanoparticles (O-$CaCO_3$)

The nanoparticles were modified with oleic acid [17]. 1-Hexane (100 mL) and oleic acid (200 µL) were mixed with stirring. Then 1 g of $CaCO_3$ nanoparticles was added to the solution at 60 °C with vigorous stirring during 5 h. The nanoparticles were then filtered, washed with ethanol, and vacuum-dried at 100 °C during 24 h [18].

2.4. Low-Density Polyethylene (LDPE)/$CaCO_3$ and LDPE/O-$CaCO_3$ Nanocomposite

The nanocomposites were prepared using a Brabender Plasti-Corder (Duisburg, Germany) internal mixer at 150 °C and a speed of 110 rpm, during 10 min. The nanocomposites with 3, 5, and 8 wt% of $CaCO_3$ nanoparticles were obtained by mixing predetermined amounts of the $CaCO_3$ as filler and neat LDPE under a nitrogen atmosphere. The samples were press-molded at 190 °C at a pressure of 50 bar during 3 min and cooled under pressure by flushing the press with cold water.

2.5. Nanoparticles and Composite Characterization

The morphology of the $CaCO_3$ was analyzed by transmission electron microscopy (TEM) (JEOL ARM 200 F, Boston, MA, USA) operating at 20 kV. Samples for TEM measurements were prepared by placing a drop of $CaCO_3$ nanoparticles on a carbon-coated standard copper grid (400 mesh).

The X-ray diffraction (XRD) patterns of the $CaCO_3$ nanoparticles were studied on a Siemens D5000 diffractometer (Berlin, Germany), using Ni-filtered Cu Kα radiation (λ = 0.154 nm). The diffraction patterns were recorded in the 2θ = 5–80° range.

FTIR measurements of $CaCO_3$ and modified nanoparticles (O-$CaCO_3$) were performed in a Bruker Vector 22 FTIR spectrometer (Karlsruhe, Germany). The infrared (IR) spectra were collected in the 4000 to 500 cm^{-1} range, with a resolution of 4 cm^{-1} at room temperature.

The tensile properties of the neat polyethylene (neat LDPE) and composites (LDPE/$CaCO_3$ and LDPE/O-$CaCO_3$) were determined on an HP model D-500 dynamometer (Buenos Aires, Argentina). The materials were molded for 3 min in a hydraulic press, HP Industrial Instruments, at a pressure of 50 bar and a temperature of 170 °C, and then cooled under pressure with water circulation. Films around 0.05 mm thickness were obtained. Dumbbell-shaped samples with an effective length of 30 mm and a width of 5 mm were cut from the compression-molded sheets. The samples were tested at a rate of 50 mm/min at 20 °C. Each set of measurements was repeated at least four times.

2.5.1. Photo-Exposure

Photoaging

Polymer films of 0.02 mm of thickness and the dimensions and 4 cm were irradiated using a Microscal Light exposure unit and Suntest/Atlas XLS 2200 W (Linsengericht, Germany) using a solar standard filter (borosilicate), which provides 550 W m^{-2} (Irradiance acc. ISO 4892/DIN 53387) in the 300–800 nm wavelength region. The temperature was kept constant at 45 °C during the testing. Exposed samples of 1 cm × 1 cm were periodically taken out and characterized. The irradiation side of the sample was alternated every 3 days. At different aging times the oxidation rates were determined on an FTIR spectrometer using the standard carbonyl index method. FTIR spectra were obtained on a Perkin Elmer BX-FTIR (Waltham, MA, USA). The polymer degradation was determined using the carbonyl index (CI) as the ratio of the optical density of the ketone carbonyl absorptions bands at 1715 cm^{-1} and the optical density corresponding to CH_2 scissoring peak at 1465 cm^{-1} [19].

Differential scanning calorimetry (DSC) was studied on a METTLER DSC823 (Columbus, OH, USA) The melting temperature and enthalpy of fusion of the neat and nanocomposite samples were determined before and after photoaging. The measurements were made at a heating rate of 10 °C·min^{-1} in an inert atmosphere. The samples were heated from 25 °C to 180 °C and then cooled to 25 °C at the same rate. Percent crystallinity (χ_c) was determined using Equation (1):

$$\chi c = \frac{\Delta H_l}{(1 - \Phi)\Delta H_0} \times 100 \quad (1)$$

where ΔH_l is the melting enthalpy (J g^{-1}) of the polymer nanocomposite, ΔH_0 is the enthalpy corresponding to the melting of a 100% crystalline sample (289 J g^{-1}) [20], and Φ is the weight fraction of the filler in the nanocomposit. The standard deviation of the T_c and T_m measurements was ca. ±2 °C.

Thermogravimetric analysis (TGA) experiments were performed on a Netzsch TG 209 F1 Libra Instrument (Selb, Germany). The films were heated from 25 °C to 600 °C at a rate of 10 °C·min^{-1} and the nitrogen flow was kept constant at 60 mL·min^{-1}. The TGA analysis also verified the content of $CaCO_3$ in the LDPE/$CaCO_3$ nanocomposites. The LDPE/$CaCO_3$ nanocomposites with 5 wt% showed a 4.65 wt% of the $CaCO_3$ nanoparticle content after the melting process.

Viscosimetric analysis before and after irradiations was carried out in o-dichlorobenzene at 135 °C in a Viscosimatic-Sofica viscometer (Santiago, Chile)

The surface morphology of the polymers before and after photoaging was characterized by scanning electron microscopy (SEM) using a Philips XL30 model instrument (Billerica, MA, USA).

3. Results and Discussion

3.1. Nanoparticle Characterization

The morphology of the nanoparticles was studied by TEM as shown in Figure 1. The nanoparticles synthesized by the coprecipitation method have an average diameter of ca. 60 nm and irregular morphology. The yield of this method was ca. 75%. The crystalline phases of the $CaCO_3$ nanoparticles were studied by XRD (Figure 2). Nanoparticles have two characteristic phases as concluded by analyzing the Bragg reflections: calcite, associated with peaks at 29° and 32° from the (104) and (006) crystal planes, respectively; and aragonite, with peaks at 27°, 30°, and 45° from the (111), (021), and (221) planes, respectively [21].

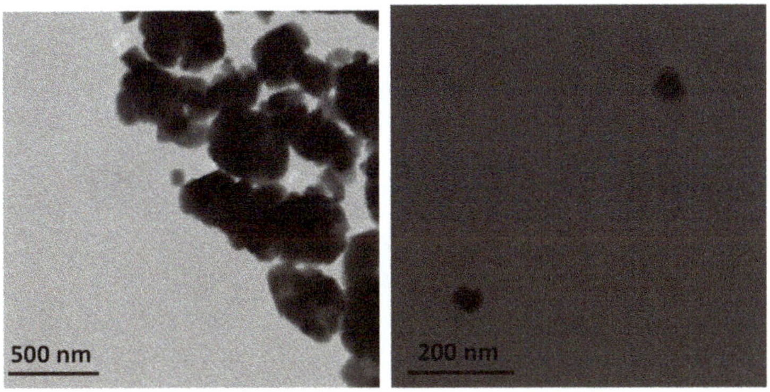

Figure 1. Transmission electron microscopy (TEM) image of the $CaCO_3$ nanoparticles.

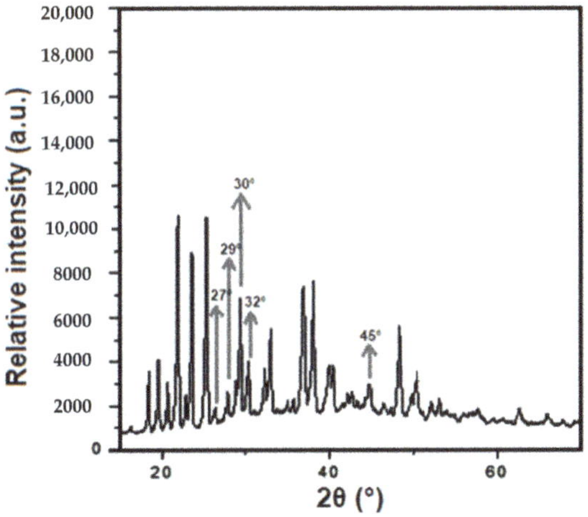

Figure 2. X-ray diffraction (XRD) spectra of $CaCO_3$ nanoparticles.

The FTIR spectrum of calcined samples ($CaCO_3$) (Figure 3b), shows the presence of calcium carbonate ($CaCO_3$) bands at 715, 880, 1490, 1804, 2530, 2900, and 2998 cm^{-1}. The band at 1490 cm^{-1} correspond to essentially asymmetric and symmetric lengthening of the O–C–O bond. Absorption bands centered at 715, 880 and 1490 cm^{-1} are characteristics of the calcite phase of $CaCO_3$ [22]. The method used for the organic modification of nanoparticles was based on that reported by Li and Zhu [17] and it was verified by FTIR spectra, where the peaks corresponding to the alkyl chain (CH_2) of oleic acid appear at 2920 cm^{-1} and 2855 cm^{-1} (Figure 3). Moreover, a small spectral line at 1710 cm^{-1} corresponding to the stretching of the carbonyl group of oleic acid indicates that the carboxylic acid group of oleic acid, −COOH, reacted with surface hydroxyl groups from the starch nanoparticles [23]. Other peaks at 1590 cm^{-1} due to carboxylate groups, and the peaks at 1550 cm^{-1} and 1430 cm^{-1} that indicate the presence of COO−, are overlapped with the characteristic band at 1490 cm^{-1} of the O–C–O bond and calcite vibrations.

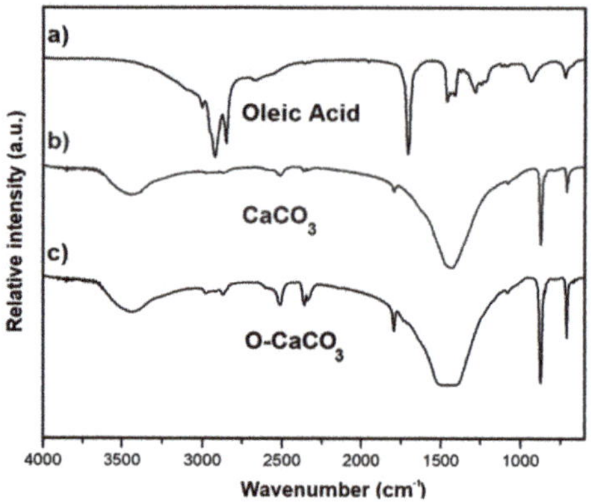

Figure 3. Fourier transform infrared (FTIR) spectra of (**a**) Oleic acid, (**b**) $CaCO_3$ nanoparticles, and (**c**) nanoparticles modified with oleic acid (O-$CaCO_3$).

3.2. Composite Characterization

3.2.1. Thermal Properties

The crystallization temperature (T_c), melting temperature (T_m), and degree of cristallinity (χ_c) were analyzed by DSC and the thermal stability obtained by TGA of the neat LDPE and LDPE/$CaCO_3$ nanocomposites are shown in Table 1. The crystallization temperature, melting temperature, and degree of crystallinity (χ_c) did not change with the incorporation of the nanoparticles, meaning that the presence of these nanoparticles did not affect the crystallization process of the polymer matrix. Similar results have been reported by other authors when different nanoparticles like ZnO, clay, silica, silver, and TiO_2 were incorporated into LDPE. This behavior may be correlated to the minimal volume fraction of the nanoparticles incorporated into the composite [24–26]. Also, the similar thermal properties of the matrix and composites would suggest analogous processing conditions as that of LDPE at a hypothetical industrial-scale production of these nanocomposites.

In the initial degradation step of the decomposition temperature, at 2% weight loss (T_2), the nanocomposites (LDPE/$CaCO_3$) were slightly more stable than LDPE nanocomposites at ca. 5%. For 10% weight loss (T_{10}), for 50% weight loss (T_{50}), and the temperature for the maximum rate of weight loss (T_{max}) did not change with the nanoparticle incorporation compared to the pure neat LDPE under inert conditions. It is well known that the incorporation of different kinds of nanofillers into a polymer can act as a superior insulator and mass transport barrier for the volatile products generated during decomposition, increasing the thermal degradation temperatures. However, these processes are relevant for high aspect ratio nanoparticles such as layered clays. Spherical-like particles with low aspect ratio should not trigger these mechanisms and only an adsorption process can explain changes in the degradation, as reported by our group on spherical silica nanoparticles [27]. In our case, the spherical-like $CaCO_3$ nanoparticles were not able to disrupt the diffusion nor adsorb volatile compounds and, therefore, no changes were observed in TGA analysis.

Table 1. Thermal properties of polyethylene (PE)/$CaCO_3$ nanocomposites before photoaging.

Process	$CaCO_3$ (wt%)	η (dL/g)	T_c (°C)	T_m (°C)	χ_c (%)	T_2 (°C)	T_{10} (°C)	T_{50} (°C)	T_{max} (°C)
Neat low-density polyethylene (LDPE)	N/A	0.44	100	112	37	385	421	459	464
LDPE/$CaCO_3$	5	0.46	100	111	37	396	421	456	460
LDPE/O-$CaCO_3$	5	0.46	101	111	35	404	420	456	462

T_c: Crystallization temperature, η: Viscosity, T_m = melting temperature; χ_c = percent crystallinity, T_2 = decomposition temperature at 2% weight loss; T_{10} = decomposition temperature at 10% weight loss; T_{50} = decomposition temperature at 50% weight loss, T_{max} = temperature for the maximum rate of weight loss (T_{max}); O-$CaCO_3$ = modified nanoparticles. The standard deviation of the viscosity measurements is ±0.03 dLg^{-1}. The standard deviation of the T_m and T_c measurements are ca. ±2 °C. The thermogravimetric analysis (TGA) has a standard deviation of ca. ±2 °C.

3.2.2. Mechanical Properties

The mechanical properties of the neat LDPE and the LDPE/$CaCO_3$ nanocomposites are displayed in Table 2 and Figure 4. An increase of Young's modulus results from adding the $CaCO_3$ nanoparticles in comparison with the neat LDPE. This performance was more pronounced with 5 wt% of nanoparticles for the LDPE/O-$CaCO_3$ nanocomposites, as Young's modulus increased ca. 29% compared to neat LDPE. Morreale et al. [3] found that 10 wt% of $CaCO_3$ fillers (50–100 nm) improved Young's modulus just ca. 20% compared to neat LDPE, due to the presence of the nanoparticle agglomeration. The increase in the modulus in our case must be caused by the strong interaction between the polymer and the nanoparticles, improving the dispersion of the particles [1]. Similar results were found by Lapcík et al. [2], who stated that due to the rigidity of the filler particles and the interaction of the filler with the polymer matrix, a reinforcement improvement can be obtained. The yield stress remained unaffected with the addition of $CaCO_3$ to neat polyethylene. A similar behavior was found for nanocomposites based on high-density polyethylene with $CaCO_3$ nanoparticles (ca. 60 nm) [9].

On the other hand, the deformation at break decreased when 5 wt% of the $CaCO_3$ nanoparticles were incorporated, probably due to many defects in the polymer matrix leading to ductility reduction [10]. The decrease of the deformation at break of LDPE/O-$CaCO_3$ (5 wt%) was slightly lower, and this may be due to the modifier improving the interaction between nanoparticles and LDPE [1].

Table 2. Mechanical properties of LDPE and LDPE/$CaCO_3$ nanocomposites.

Process	$CaCO_3$ Content (wt%)	E (MPa)	σy (MPa)	ε_{Break} (%)
Neat LDPE	0	202 ± 7	8.5 ± 0.03	70.3 ± 8
LDPE/$CaCO_3$	3	230 ± 7	9.1 ± 0.15	62.5 ± 10
	5	250 ± 4	9.2 ± 0.09	39.8 ± 3
LDPE/O-$CaCO_3$	5	260 ± 10	8.8 ± 0.14	61.1 ± 1

E = Young's modulus; σy = yield stress; ε_{Break} = deformation at break.

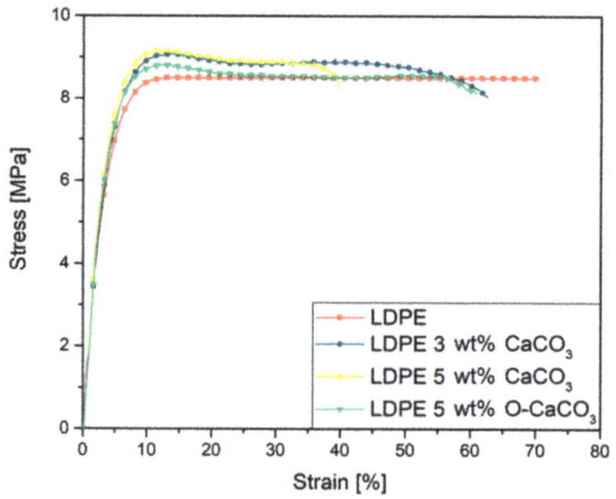

Figure 4. Stress-strain curves for neat LDPE, LDPE/$CaCO_3$ with 3 and 5 wt%, and LDPE/O-$CaCO_3$ with 5 wt% nanocomposites.

3.3. *Photoaging Analysis*

3.3.1. Thermal and Mechanical Properties

Crystallization temperature (T_c), melting temperature (T_m), degree of cristallinity (χ_c), thermal stability analysis, and viscosity of the neat LPE and LDPE/$CaCO_3$ nanocomposites after photoaging are displayed in Table 3. After photoaging, the χ_c for nanocomposites increased slightly compared to pure LDPE and LLDPE/$CaCO_3$ before irradiation. This behavior has been attributed to recrystallization due to LDPE scission of end chains producing mobile small chain fragments able to undergo reorganization and recrystallization [14,28]. This scission is confirmed also finding that after photoaging the viscosity decreased due to the formation of low molecular weight compounds during aging (Table 3). It should be noted that this behavior is slightly greater for nanocomposites than for neat LDPE. These results show that the incorporation of nanoparticles into the polymer accelerates its degradation. After photoaging, both decomposition temperatures, (T_{10}) and T_{max}, did not change. In previous work using Ca and Fe stereates as PE degradant, the authors found a slight decrease in T_{10}, and they explained this behavior as due to the prooxidative nature of stereate during the photoaging process [29].

Table 3. Thermal properties of PE/$CaCO_3$ nanocomposites after photoaging.

		Photoaging					
Nanoparticles	**$CaCO_3$ (wt%)**	**η (dL/g)**	**T_c (°C)**	**T_m (°C)**	**χ_c**	**T_{10} (°C)**	**T_{max} (°C)**
Neat LDPE	N/A	0.18	105	107	40	423	467
LDPE/$CaCO_3$	5	0.13	106	108	44	426	460

T_c: crystallization temperature, T_m = melting temperature; χ_c = percent crystallinity. T_{10} = decomposition temperature at 10% weight loss; T_{max} = temperature for the maximum rate of weight loss (T_{max}). Photoaging during 10 days.

The mechanical properties were evaluated after 10 days of photoaging as displayed in Table 4. The polymers were difficult to break into pieces by hand after 10 days of irradiation, confirming the strong degradation. Young's modulus for the LDPE/$CaCO_3$ increased after photoaging compared to photoaged neat PE. Young's modulus increased after photo-oxidation mainly due to a significant embrittlement of the material and recrystallization phenomena caused by scission reactions [15]. These results further confirmed that nanoparticles accelerated the degradation of the polymer.

Table 4. Mechanical properties of LDPE and LDPE/$CaCO_3$ nanocomposites after irradiation during 10 days.

		Photoaging		
Nanoparticles	**$CaCO_3$ (wt%)**	**E (MPa)**	**σy (MPa)**	**ε_{Break} (%)**
Neat LDPE	N/A	214 ± 14	7 ± 0.3	20 ± 4
LDPE/$CaCO_3$	5	301 ± 15	6 ± 0.3	8 ± 3

E = Young's modulus; σy = yield stress; ε_{Break} = deformation at break. The photoaging during 10 days.

3.3.2. Carbonyl Index

The degradation, calculated by FTIR, and the carbonyl index (CI) after 35 days of irradiation are displayed in Figure 5. The carbonyl index was measured for neat LDPE and the LDPE/$CaCO_3$ sample with 5 wt% of nanoparticles. The LDPE/$CaCO_3$ nanocomposite with 5 wt% had a higher carbonyl index than LDPE, showing an influence in the degradation of the polymers with nanoparticle incorporation. The IR spectra of the photoaged polymer (LDPE/$CaCO_3$) has a strong peak at 1720 cm^{-1}, which is related to the C=O stretching vibration of the carbonyl group (Figure 6) [30]. The second band around 3400 cm^{-1} is related to the hydroxyl group, which indicates the generation of hydroperoxides and hydroxyl species. Furthermore, the intensity of the carbonyl and hydroxyl bands grew with increasing exposure time [14]. The intensity peaks at 2930, 2850, 1470, and 720 cm^{-1}, corresponding to the alkyl chain, decreased slightly. Carboxylic acid salts can be formed by the reaction between the carboxylic acids coming from photoaged PE and the basic fillers. Li et al. [14] explained that PP/$CaCO_3$ shows a higher degradation rate than neat PP, due to functional groups on the surface of the nanoparticles catalyzing the photooxidant ion reaction of PP. There are absorbed hydroxyl groups on the surface of the nanofillers, which are active in photo-chemical reactions. Therefore, the hydrophilic surface of the nanoparticles is responsible for the increased polymer degradation. Further degradation in the abiotic environment is through the Norrish type I and II mechanism, giving rise to esters and ketones [31].

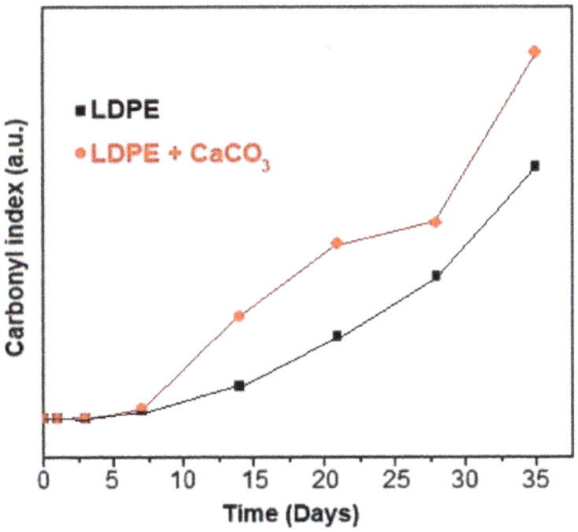

Figure 5. Carbonyl index (CI) of neat LDPE and LDPE/$CaCO_3$ at different irradiation times.

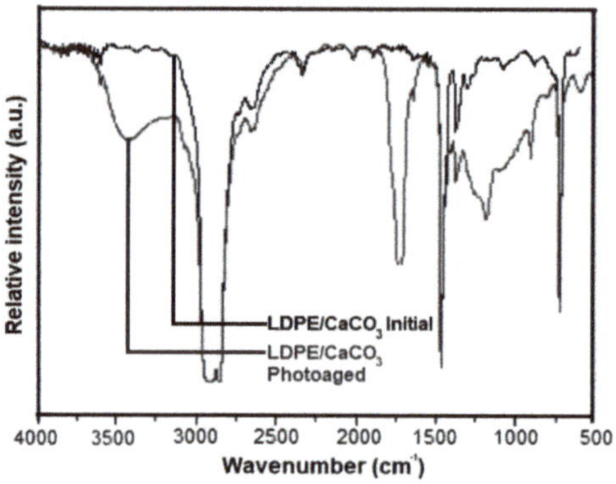

Figure 6. Infrared (IR) spectra of initial LDPE/$CaCO_3$ nanocomposites and LDPE/$CaCO_3$ after photoaging for 35 days.

SEM images of LDPE and LDPE/$CaCO_3$ with 5 wt% of $CaCO_3$ before and after photoaging are shown in Figure 7. Before irradiation, LDPE and LDPE/$CaCO_3$ images exhibit a smooth and homogeneous surface morphology (Figure 7a,c). After irradiation, the morphology is changed, the LDPE and LDPE/$CaCO_3$ nanocomposites presented some cavities, with nanoparticles producing larger ones (Figure 7b,d). After irradiation, the nanocomposites undergo greater acceleration of photodegradation than neat LDPE, confirming the results shown above by CI, mechanical properties, and viscosity.

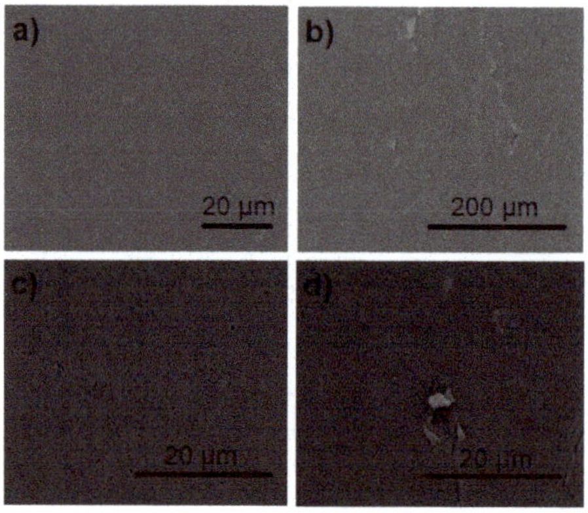

Figure 7. Scanning electron microscopy (SEM) images of initial and photoaged PE and PE/SNp during 35 days of photoaging: (**a**) PE initial; (**b**) PE aged; (**c**) PE/$CaCO_3$ initial; and (**d**) PE/$CaCO_3$ aged.

4. Conclusions

The co-precipitation method was used to produce $CaCO_3$ (60 nm), which were then modified organically with oleic acid (O-$CaCO_3$). Young's modulus increased ca. 29% for LDPE/O-$CaCO_3$ compared to the neat LDPE.

Regarding polymer photoaging, the degree of crystallinity (χ_c) increased with photoaging, and this effect was higher for LDPE/$CaCO_3$ (ca. 19%) nanocomposites than for neat LDPE (ca. 8%), attributed to recrystallization of the polymer. The viscosity of LDPE decreased by ca. 59% after photoaging and around 72% for LDPE/$CaCO_3$, as indicated by the decreased molecular weight of the polymer due to chain scissions, and the pronounced effect of the nanoparticles in the polymer degradation. Young's modulus increased ca. 16% for LDPE/O-$CaCO_3$ after photoaging because the nanoparticles accelerate the polymer's degradation. The degradation of the films obtained was confirmed by the carbonyl index, where carbonyl bands appear more intense. LDPE/$CaCO_3$ with 5 wt% had a high carbonyl index, showing an influence in the degradation of the polymers with the incorporation of nanoparticles.

Author Contributions: Conceptualization, B.D. and A.A.; Methodology, M.P.R.; Validation, J.A.O.; Formal Analysis, F.S. and C.O.; Investigation, P.A.Z. and H.P.; Resources, P.A.Z.; Data Curation, F.S. and C.O.; Writing-Original Draft Preparation, P.A.Z.; Writing-Review & Editing, P.A.Z., F.S., H.P.; Visualization, F.S.; Supervision, P.A.Z. and H.P.; Project Administration, P.A.Z.; Funding Acquisition, P.A.Z.

Funding: This research was funded by [FONDECYT Regular Project] grant number [1170226] and FIA, [Fundación para la Innovación Agraria] under FIA project "FIA-PYT-2013-0018" (http://www.fia.cl/); and "Gobierno Regional Metropolitano de Santiago" (GORE-RM). P.A.Z. acknowledges the financial support of Project [DICYT] grant number [051641ZR_DAS], Vicerrectoria de Investigación, Desarrollo e Innovación, Universidad de Santiago de Chile. H.P. acknowledges the financial support of the [FONDECYT Project] grant number [1150130].

Acknowledgments: P.A.Z. acknowledges the financial support under FONDECYT Regular Project 1170226; FIA, "Fundación para la Innovación Agraria" under FIA project "FIA-PYT-2013-0018" (http://www.fia.cl/); and "Gobierno Regional Metropolitano de Santiago" (GORE-RM). P.A.Z. acknowledges the financial support of Project DICYT, 051641ZR_DAS, Vicerrectoria de Investigación, Desarrollo e Innovación, Universidad de Santiago de Chile. H.P. acknowledges the financial support of the FONDECYT Project 1150130.

Conflicts of Interest: The authors declare no conflict of interest.

References

1. Chan, C.M.; Wu, J.S.; Li, J.X.; Cheung, Y.K. Polypropylene/Calcium carbonate nanocomposites. *Polym. J.* **2001**, *43*, 2981–2992. [CrossRef]
2. Lapcík, L.; Mañas, D.; Vasina, D.; Lapcíkova, B.; Reznícek, M.; Zadrapa, P. High density poly (ethylene)/$CaCO_3$ hollow spheres composites for technical applications. *Compos. Part B Eng.* **2017**, *113*, 218–224. [CrossRef]
3. La Mantia, F.P.; Morreale, M.; Scaffaro, R.; Tulone, S. Rheological and mechanical behavior of LDPE/calcium carbonate nanocomposites and microcomposites. *J. Appl. Polym. Sci.* **2013**, *127*, 2544–2552. [CrossRef]
4. Zebarjad, S.M.; Sajjadi, S.A. On the strain rate sensitivity of HDPE/$CaCO_3$ nanocomposites. *Mat. Sci. Eng. A* **2008**, *475*, 365–367. [CrossRef]
5. Wong, A.C.-Y.; Wong, A.C.M. Extrudate swell ratio characteristics of $CaCO_3$ added linear low density polyethylene. *Polym. Test.* **2018**, *71*, 262–271. [CrossRef]
6. Luo, Z.; Wang, Y.; Wang, H.; Feng, S. Impact of nano-$CaCO_3$-LDPE packaging on quality of fresh-cut sugarcane. *J. Sci. Food Agric.* **2014**, *94*, 3273–3280. [CrossRef]
7. Silapasorn, K.; Sombatsompop, K.; Kositchaiyong, A.; Wimolmala, E.; Markpin, T. Effect of chemical structure of thermoplastics on antibacterial activity and physical diffusion of triclosan doped in vinyl thermoplastics and their composites with $CaCO_3$. *J. Appl. Polym. Sci.* **2011**, *121*, 253–261. [CrossRef]
8. Lazzeri, A.; Zebarjad, S.M.; Pracella, M.; Cavalier, K.; Rosa, R. Filler toughening of plastics. Part 1—The effect of surface interactions on physico-mechanical properties and rheological behaviour of ultrafine $CaCO_3$/HDPE nanocomposites. *Polym. J.* **2005**, *46*, 827–844. [CrossRef]
9. Deshmane, C.; Yuan, Q.; Misra, R.D.K. On the fracture characteristics of impact tested high density polyethylene–calcium carbonate nanocomposites. *Mat. Sci. Eng. A* **2007**, *452–453*, 592–601. [CrossRef]
10. Pradittham, A.; Charitngam, C.; Puttajan, S.; Atong, D.; Pechyen, C. Surface modified $CaCO_3$ by palmitic acid as nucleating agents for polypropylene film: Mechanical, thermal and physical properties. *Energy Procedia* **2014**, *56*, 264–273. [CrossRef]
11. Lin, Y.; Chen, H.; Chan, C.; Wu, J. Effects of coating amount and particle concentration on the impact toughness of polypropylene/$CaCO_3$ nanocomposites. *Eur. Pol. J.* **2011**, *47*, 294–304. [CrossRef]
12. Zhang, Q.Z.; Yu, Z.; Xie, X.; Mai, Y.W. Crystallization and impact energy of polypropylene/$CaCO_3$ nanocomposites with nonionic modifier. *Polym. J.* **2004**, *45*, 5985–5994. [CrossRef]
13. Kyzioł, K.; Oczkowska, J.; Kottfer, D.; Klich, M.; Kaczmarek, L.; Kyzioł, A.; Grzesik, Z. Physicochemical and biological activity analysis of low-density polyethylene substrate modified by multi-layer coatings based on DLC structures, obtained using RF CVD method. *Coatings* **2018**, *8*, 135. [CrossRef]
14. Li, J.; Yang, R.; Yu, J.; Liu, Y. Natural photo-aging degradation of polypropylene. *Polym. Degrad. Stab.* **2008**, *93*, 84–89. [CrossRef]

15. Morreale, M.; Dintcheva, N.T.; La Mantia, F.P. The role of filler type in the photo-oxidation behaviour of micro- and nano-filled polypropylene. *Polym. Int.* **2011**, *60*, 1107–1116. [CrossRef]
16. Babou-Kammoe, R.; Hamoudi, S.; Larachi, F.; Belkacem, K. Synthesis of $CaCO_3$ nanoparticles by controlled precipitation of saturated carbonate and calcium nitrate aqueous solutions. *Can. J. Chem Eng.* **2012**, *90*, 26–33. [CrossRef]
17. Li, Z.; Zhu, Y. Surface-modification of SiO_2 nanoparticles with oleic acid. *Appl. Surf. Sci.* **2003**, *211*, 315–320. [CrossRef]
18. Yañez, D.; Rabagliati, F.M.; Guerrero, S.; Lieberwirth, I.; Ulloa, M.T.; Gomez, T.; Zapata, P. Photocatalytic inhibition of bacteria by TiO_2/nanotubes-doped polyethylene composites. *Appl. Catal. A* **2015**, *489*, 255–261. [CrossRef]
19. Zapata, P.A.; Rabagliati, F.M.; Lieberwirth, I.; Catalina, F.; Corrales, T. Study of the photodegradation of nanocomposites containing TiO_2 nanoparticles dispersed in polyethylene and in poly(ethylene-co-octadecene). *Polym. Degrad. Stab.* **2014**, *109*, 106–114. [CrossRef]
20. Wei, L.; Tang, T.; Huang, B. Synthesis and characterization of polyethylene/clay–silica nanocomposites: A montmorillonite/silica-hybrid-supported catalyst and in situ polymerization. *J. Polym. Sci. A Polym. Chem.* **2004**, *42*, 941–949. [CrossRef]
21. Casanova, H.; Higuita, L.P. Synthesis of calcium carbonate nanoparticles by reactive precipitation using a high pressure jet homogenizer. *Chem. Eng. J.* **2011**, *175*, 569–578. [CrossRef]
22. Ghiasi, M.; Malekzadeh, A. Synthesis of $CaCO_3$ nanoparticles via citrate method and sequential preparation of CaO and $Ca(OH)_2$ nanoparticles. *Cryst. Res. Technol.* **2012**, *47*, 471–478. [CrossRef]
23. Gao, Y.; Chen, G.; Oli, Y.; Zhang, z.; Xue, Q. Study on tribological properties of oleic acid-modified TiO_2 nanoparticle in water. *Wear* **2002**, *252*, 454–458. [CrossRef]
24. Redhwi, H.H.; Siddiqui, M.N.; Andrady, A.L.; Syed, H. Durability of LDPE nanocomposites with clay, silica, and zinc oxide—Part I: Mechanical properties of the nanocomposite materials. *Polym. Compos.* **2013**, *34*, 1878–1883. [CrossRef]
25. Zapata, P.A.; Palza, H.; Cruz, L.S.; Lieberwirth, I.; Catalina, F.; Corrales, T.; Rabagliati, F.M. Polyethylene and poly(ethylene-co-1-octadecene) composites with TiO_2 based nanoparticles by metallocenic "in situ" polymerization. *Polym. J.* **2013**, *54*, 2690–2698. [CrossRef]
26. Zapata, P.A.; Tamayo, L.; Páez, M.; Cerda, E.; Azócar, I.; Rabagliati, F.M. Nanocomposites based on polyethylene and nanosilver particles produced by metallocenic "in situ" polymerization: Synthesis, characterization, and antimicrobial behavior: Synthesis, characterization, and antimicrobial behavior. *Eur. Polym. J.* **2011**, *47*, 1541–1549. [CrossRef]
27. Palza, H.; Vergara, R.; Zapata, P.A. Composites of polypropylene melt blended with synthesized silica nanoparticles. *Compos. Sci. Technol.* **2011**, *71*, 535–540. [CrossRef]
28. Hsu, Y.; Weir, M.; Truss, R.; Garvey, C.; Nicholson, T.; Halley, P. A fundamental study on photo-oxidative degradation of linear low-density polyethylene films at embrittlement. *Polym. J.* **2012**, *53*, 2385–2393. [CrossRef]
29. Pablos, J.L.; Abrusci, C.; Marín, I.; Lopez-Marín, J.; Catalina, F.; Espí, E.; Corrales, T. Photodegradation of polyethylenes: Comparative effect of Fe and Ca-stearates as pro-oxidant additives. *Polym. Degrad. Stab.* **2010**, *95*, 2057–2064. [CrossRef]
30. Thomas, R.A.; Nair, V.; Sandhyarani, N. TiO_2 nanoparticle assisted solid phase photocatalytic degradation of polythene film: A mechanistic investigation. *Colloids Surf. A Physicochem. Eng. Asp.* **2013**, *422*, 1–9. [CrossRef]
31. Wiles, D.M.; Scott, G. Polyolefins with controlled environmental degradability. *Polym. Degrad. Stab.* **2006**, *91*, 1581–1592. [CrossRef]

Sample Availability: Samples of the compounds are not available from the authors.

molecules

MDPI

Article

Fabrication of Spherical Titania Inverse Opal Structures Using Electro-Hydrodynamic Atomization

Jong-Min Lim * **and Sehee Jeong**

Department of Chemical Engineering, Soonchunhyang University, 22 Soonchunhyang-ro, Shinchang-myeon, Asan-si, Chungcheongnam-do 31538, Korea; jsh951004@naver.com

* Correspondence: jmlim@sch.ac.kr; Tel.: +82-41-530-4961

Academic Editors: Marinella Striccoli, Roberto Comparelli and Annamaria Panniello

Received: 29 September 2019; Accepted: 26 October 2019; Published: 30 October 2019

Abstract: Spherical PS/HEMA opal structure and spherical titania inverse opal structure were fabricated by self-assembly of colloidal nanoparticles in uniform aerosol droplets generated with electro-hydrodynamic atomization method. When a solution of PS/HEMA nanoparticles with uniform size distribution was used, PS/HEMA nanoparticles self-assembled into a face-centered cubic (FCC) structure by capillary force with the evaporation of the solvent in aerosol droplet, resulting in a spherical opal structure. When PS/HEMA nanoparticles and anatase titania nanoparticles were dispersed simultaneously into the solution, titania nanoparticles with relatively smaller size were assembled at the interstitial site of PS/HEMA nanoparticles packed in the FCC structure, resulting in a spherical opal composite structure. Spherical titania inverse opal structure was fabricated after removing PS/HEMA nanoparticles from the spherical opal composite structure by calcination.

Keywords: colloidal crystal; inverse opal; electro-hydrodynamic atomization; photonic ball; titania

1. Introduction

Photonic bandgap is a phenomenon that originated from periodically arranged materials having different refractive indices. It can be controlled by changing the structure of periodicity and the refractive index of constituent material. Yablonovitch et al. have demonstrated the fabrication of 3D photonic crystal having a photonic bandgap in the microwave region [1]. Since then, there have been various studies on the fabrication of 3D photonic crystal structures [2–7]. With a bottom-up approach, colloidal self-assembled structures have been widely adopted to make 3D photonic crystal structures due to their simplicity and cost-effectiveness. As the solvent of colloidal dispersion evaporates slowly, nanoparticles with uniform size distribution are self-assembled into a hexagonal crystal lattice (i.e., face-centered cubic (FCC) structure) to form an opal structure. In addition, an inverse opal structure with more robust photonic bandgap can be prepared by filling interstitial sites of the opal structure using materials with high refractive index and removing colloids having a hexagonal crystal lattice [3,5,6,8].

One of the main issues in the fabrication of 3D colloidal photonic crystals (e.g., opal and inverse opal structures) is the control of their shape and size in a reproducible manner. In order to control size and shape, various studies have prepared spherical 3D colloidal crystals using uniform-sized droplets as confined geometries [6–9]. Electro-hydrodynamic atomization method, also known as electrospray, has been developed for large-scale production of spherical 3D colloidal crystals. Moon et al. have demonstrated electro-hydrodynamic atomization for large-scale production of spherical polystyrene opal structures and spherical silica inverse opal structures [10]. Since they used water as a solvent, an additional process was required for the evaporation of the solvent. In addition, spherical titania inverse opal structures could not be fabricated due to poor dispersion stability of titania nanoparticles in aqueous solution. Hong et al. have prepared spherical silica opal structures without an additional solvent evaporation process using ethanol as a solvent in electro-hydrodynamic atomization [11]. They also used crosslinked polystyrene nanoparticles in toluene for electro-hydrodynamic atomization. However, spherical structures with irregularly packed colloidal crystal shells and hollow cores were fabricated due to the extremely fast evaporation rate of toluene.

In this study, we used poly [styrene-co-(2-hydroxyethyl methacrylate)] nanoparticles (PS/HEMA nanoparticles) with enhanced stability in ethanol to fabricate spherical PS/HEMA opal structures and spherical titania inverse opal structures by electro-hydrodynamic atomization. Without the additional evaporation process, we could prepare compact spherical PS/HEMA opal structures regularly packed in an FCC structure. When a mixture of PS/HEMA nanoparticles in ethanol and anatase titania nanoparticles in methanol is used for electro-hydrodynamic atomization, titania nanoparticles of relatively small size were assembled at the interstitial site of PS/HEMA nanoparticles packed in the FCC structure, resulting in a spherical opal composite structure. After calcination to remove PS/HEMA nanoparticles, spherical titania inverse opal could be fabricated. Since the electro-hydrodynamic atomization method can rapidly prepare a large amount of spherical PS/HEMA opal structures and spherical titania inverse opal structures, it can expedite the commercial application of spherical opal and spherical inverse opal structures in various areas, including reflective mode display, photo catalysis, solar cell electrode materials, and analytical systems [12–17].

2. Results

2.1. Characterization of Monodisperse PS/HEMA Nanoparticles and Titania Nanoparticles

Figure 1a shows an SEM image of monodisperse PS/HEMA nanoparticles. PS/HEMA nanoparticles were hexagonally packed (i.e., FCC structure), confirming uniform size distribution of PS/HEMA nanoparticles. Figure 1b shows a TEM image of titania nanoparticles. These titania nanoparticles had an anatase phase based on power x-ray diffraction as in Figure 1c.

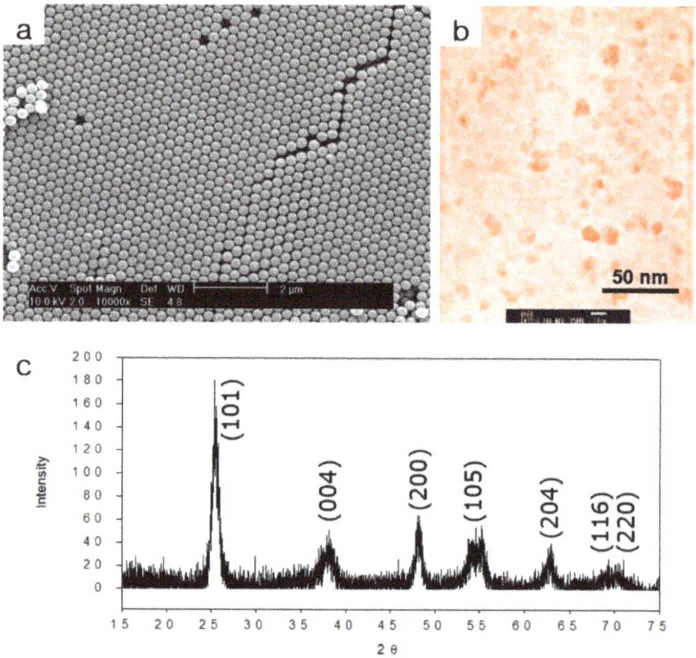

Figure 1. (**a**) SEM image of PS/HEMA latex nanoparticles. (**b**) TEM image and (**c**) x-ray diffraction data of titania colloidal nanoparticles.

2.2. Preparation of Uniform Aerosol Droplets Using Electro-hydrodynamic Atomization

AC electric field in the range of 1.2–1.8 kV/mm intensity with frequency of 1–5 kHz was used to maintain stable Taylor cone jet mode. The frequency of the AC electric field did not significantly affect the size of aerosol droplets because the frequency range (i.e., 1–5 kHz) was high enough [11].

2.3. Fabrication of Spherical PS/HEMA Opal Structures Structures

Figures 2a–c and 2d–f show SEM and digital camera images of the spherical opal structure consisting of 280 and 210-nm-sized monodisperse PS/HEMA nanoparticles, respectively. As shown in Figure 2a,d, spherical opal structure could be fabricated as the solvent evaporated. As shown in magnified SEM images (Figure 2b,e), monodisperse PS/HEMA nanoparticles self-assembled into a hexagonal crystal lattice, showing the (111) plane of the FCC structure.

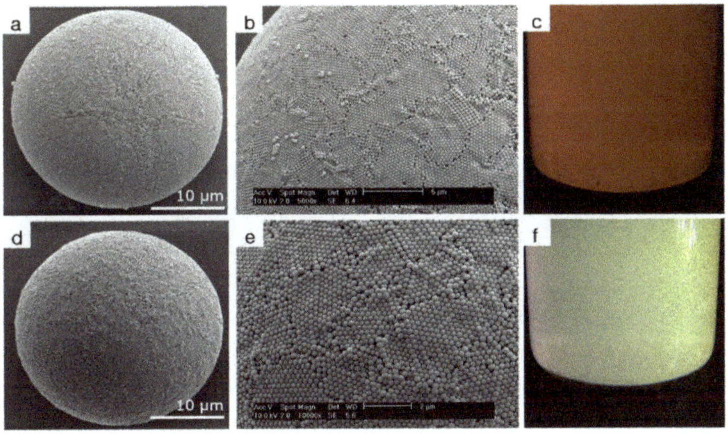

Figure 2. (**a**,**b**) SEM images and (**c**) digital camera image of spherical opal structures made of 280 nm-sized PS/HEMA nanoparticles. (**d**,**e**) SEM images and (**f**) digital camera image of spherical opal structures from 210-nm-sized PS/HEMA nanoparticles.

2.4. Fabrication of Spherical Titania Inverse Opal Structures

Since both PS/HEMA nanoparticles and anatase titania nanoparticles could make stable colloidal dispersion in the mixture of ethanol and methanol at a 3:1 volumetric ratio, we could use mixed colloidal dispersion for electro-hydrodynamic atomization. Figure 3a,b show SEM images of spherical opal composite structures consisting of PS/HEMA nanoparticles and anatase titania nanoparticles. As shown in Figure 3c,d, spherical titania inverse opal structures could be obtained after removing PS/HEMA nanoparticles by 500 °C calcination. The inset of Figure 3b,d shows the fast Fourier transform (FTT) of the SEM images.

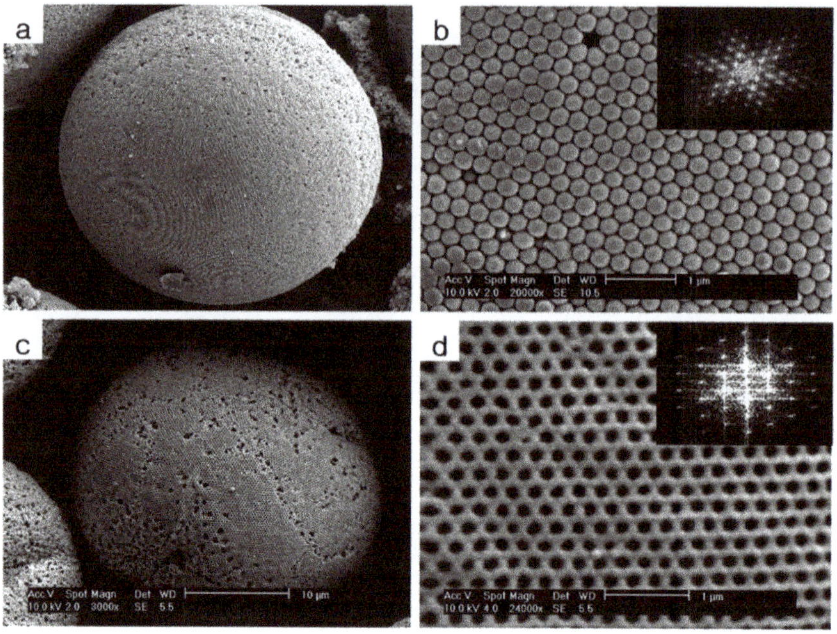

Figure 3. (**a**,**b**) SEM images of spherical opal composite structures consisting of PS/HEMA nanoparticles and titania nanoparticles. (**c**,**d**) SEM images of spherical titania inverse opal structure. Inset shows the fast Fourier transform (FTT) of the SEM images.

3. Discussion

The size of the spherical opal structure was determined by the size of the aerosol droplet produced by electro-hydrodynamic atomization and amounts of nanoparticles in an aerosol droplet. Since the concentration of nanoparticles in the solution was kept constant, uniform aerosol droplets should be stably generated to produce spherical opal structures with uniform size distribution.

Uniform aerosol droplets of several tens to several hundreds of micrometers in size could be generated in a controlled manner using electro-hydrodynamic atomization method. When colloidal dispersion was injected into the capillary needle at a constant flow rate using a syringe pump, droplets were formed at the end of the capillary. When gravity became greater than restoring surface tension, the droplet detached from the capillary and dripped through the ring electrode. In the electro-hydrodynamic atomization method, AC electric field was applied to a stainless capillary needle and a ring electrode. Because AC electric field deforming electrical tangential stress was applied to the meniscus, when the electric field increased, the size of the dripping droplet became smaller due to electrical tangential stress known as dripping mode. When the applied electric field exceeded the threshold value, the meniscus became a cone shape known as Taylor cone jet mode [18]. In the Taylor cone jet mode, droplets with uniform size distribution could be generated stably. As electric field strength further increased, multiple unstable jets were formed at the end of the stainless steel capillary known as multi-jet mode [10,11].

The solvent in aerosol droplets evaporated while droplets passed the cylindrical plastic tube. As the solvent evaporated, nanoparticles in the droplet gradually got closer and began to self-assemble by capillary force. When nanoparticles were brought into contact with each other, nanoparticles were fixed to a spherical opal structure by van der Waals force. The self-assembly and fixation process could be completed in a few seconds after droplet generation by electro-hydrodynamic atomization.

Since the sizes of titania nanoparticles were much smaller than those of PS/HEMA nanoparticles, titania nanoparticles were assembled at the interstitial site of hexagonally packed PS/HEMA nanoparticles. The periodicities of the structures in Figure 3b,d are 239 nm and 224 nm, respectively. The periodicity of the structure was decreased by about 6.3% by the calcination. Because of the defects in Figure 3b, there are slight differences between the average diameter of PS/HEMA nanoparticles and the periodicity of the structure.

A large amount of spherical PS/HEMA opal structure could be obtained rapidly using the electro-hydrodynamic atomization method. As shown in Figure 2c,f, spherical PS/HEMA opal structures were dispersed in water in order to demonstrate optical properties. Reflected diffraction colors could be controlled depending on the size of nanoparticles constituting the spherical opal structure. Since the surface of the spherical opal structure was composed of the (111) plane of the FCC structure, light corresponding to the bandgap of the photonic crystal was reflected. The bandgap position in wavelength (λ) could be estimated by Bragg's law for the (111) plane of the FCC structure: [2,19]

$$\lambda = 2dn_{eff} = \left(\frac{8}{3}\right)^{0.5} Dn_{eff} \tag{1}$$

where d was the spacing of the (111) plane, n_{eff} was the effective refractive index, and D was the diameter of constituent nanoparticles. The n_{eff} could be obtained by

$$n_{eff} = \left\{\phi_p n_p^2 + (1 - \phi_p) n_m^2\right\} \tag{2}$$

where ϕ_p was the volume fraction of nanoparticles, and n_p and n_m were refractive indices of nanoparticles and matrix, respectively. The bandgap position (λ) could be estimated from the Bragg's law using the volume fraction of nanoparticles in the FCC structure (ϕ_p = 0.74), refractive index of PS/HEMA particles (n_p = 1.59), and refractive index of water (n_m = 1.33). For spherical opal structures composed of 280 and 210-nm-sized PS/HEMA nanoparticles, bandgap positions (λ) estimated from the Bragg's law were 698 nm and 523 nm, respectively. Since spherical PS/HEMA opal structure could be fabricated in large quantities using the electro-hydrodynamic atomization method, bandgap positions (λ) can be visually confirmed from the digital camera images shown in Figure 2c,f.

The bandgap position (λ) of the spherical titania inverse opal structure could be estimated from Bragg's law using the refractive index of anatase phase titania (n_m = 2.5) and the refractive index of water (n_p = 1.33). Here, we assumed that 74% of the unit cell structure was occupied by water for the case of spherical inverse opal structure. When 280-nm-sized PS/HEMA nanoparticles were used to make the spherical titania inverse opal structure, the bandgap was located in the near infrared region (i.e., λ = 783 nm). The bandgap position could be controlled by changing the size of PS/HEMA nanoparticles used to fabricate spherical titania inverse opal structures. The bandgap position can also be controlled by changing the refractive indices of materials used to make spherical inverse opal structures and solvents used to disperse the spherical inverse opal structure.

4. Materials and Methods

4.1. Chemicals

All chemicals and solvents were reagent grade and used without further purification. Styrene (99%) was purchased from Kanto chemical (Tokyo, Japan), and 2-hydroxyethyl methacrylate (reagent grade) was purchased from Sigma-Aldrich (St. Louis, MO, USA). Potassium persulfate (98%) as the initiator was purchased from Kanto chemical (Tokyo, Japan). Titania nanoparticles in methanol dispersion (DH 60) were obtained from Nissan chemicals (Tokyo, Japan). Ethanol (≥ 99.9%) was purchased from Merck (Kenilworth, NJ, USA).

4.2. Synthesis of Monodisperse PS/HEMA Nanoparticles

Monodisperse PS/HEMA nanoparticles were synthesized through batch type surfactant-free emulsion copolymerization in aqueous medium. Detailed synthesis procedures can be found elsewhere [20]. Aqueous PS/HEMA nanoparticle dispersion was dried on a piece of silicon wafer and coated with gold for observation using a field emission scanning electron microscope (FE-SEM, XL305FEG, Philips (Amsterdam, The Netherlands)).

4.3. Fabrication of Aerosol Droplets Using Electro-hydrodynamic Atomization

PS/HEMA nanoparticles were re-dispersed in ethanol prior to electro-hydrodynamic atomization. To make titania inverse opal, PS/HEMA in ethanol dispersion and titania in methanol dispersion were mixed at 3:1 volumetric ratio. Vortex mixer (Maxi mix II, Thermo Scientific (Waltham, MA, USA)) and ultrasonic cleaner (EW-08893-21, Cole-parmer (Vernon, IL, USA)) were used for re-dispersion and the mixing process.

Aerosol droplets with uniform size distribution were prepared by using the electro-hydrodynamic atomization method. The schematic of the experimental set up for electro-hydrodynamic atomization is illustrated in Scheme 1a. The apparatus for electro-hydrodynamic atomization consists of a colloidal dispersion injection part, an electric field control part, a meniscus observation part, and a colloidal self-assembled structure collection part. In the colloidal dispersion injection part, colloidal dispersion was introduced to a disposable 1-mL syringe which was connected to a stainless capillary needle (metal hub needle, Hamilton (Reno, NV, USA)). The disposable syringe was loaded to a syringe pump (KDS 100, KD Scientific (Holliston, MA, USA)) to maintain a constant flow rate of 0.5 mL/h. In the electric field control part, an arbitrary function generator (DS345, Stanford Research Systems (Sunnyvale, CA, USA)) and a high voltage power amplifier (20/20B, Trek Inc. (Lockport, NY, USA)) were connected to a stainless capillary needle and a ring electrode with circular hole of 1 cm in diameter. The shape of the meniscus formed on the capillary was changed with AC electric field. The AC electric field was maintained in the range of 1.2–1.8 kV/mm with 1–5 kHz frequency for stable generation of droplets in Taylor cone jet mode [18]. In order to precisely observe the shape, a CCD camera with a ring-shaped illuminator was used in the meniscus observation part. At the bottom of the ring electrode, there was a 90-cm-long cylindrical plastic tube for evaporating solvent in aerosol droplets. When the solvent in the aerosol droplet was completely evaporated as shown in Scheme 1b, PS/HEMA nanoparticles inside an aerosol droplet self-assembled to form spherical PS/HEMA opal structures. In addition, when PS/HEMA nanoparticles and titania nanoparticles were used simultaneously as shown in Scheme 1c, titania nanoparticles with smaller size were self-assembled at the interstitial site of hexagonally-packed PS/HEMA nanoparticles. As the solvent evaporated, spherical opal composite structures were formed as shown in Scheme 1c. Spherical opal structures and spherical opal composite structures could be collected by placing a glass petri-dish under the cylindrical plastic tube. As shown in Scheme 1c, spherical titania inverse opal structures could be obtained by removing PS/HEMA nanoparticles from the spherical opal composite structures via 500 °C calcination for about 8 hours using a muffle furnace (PK9712150030-9, Isuzu (Tokyo, Japan)). Spherical PS/HEMA opal structures and spherical titania inverse opal structures were observed using a FE-SEM (XL305FEG, Philips (Amsterdam, The Netherlands)).

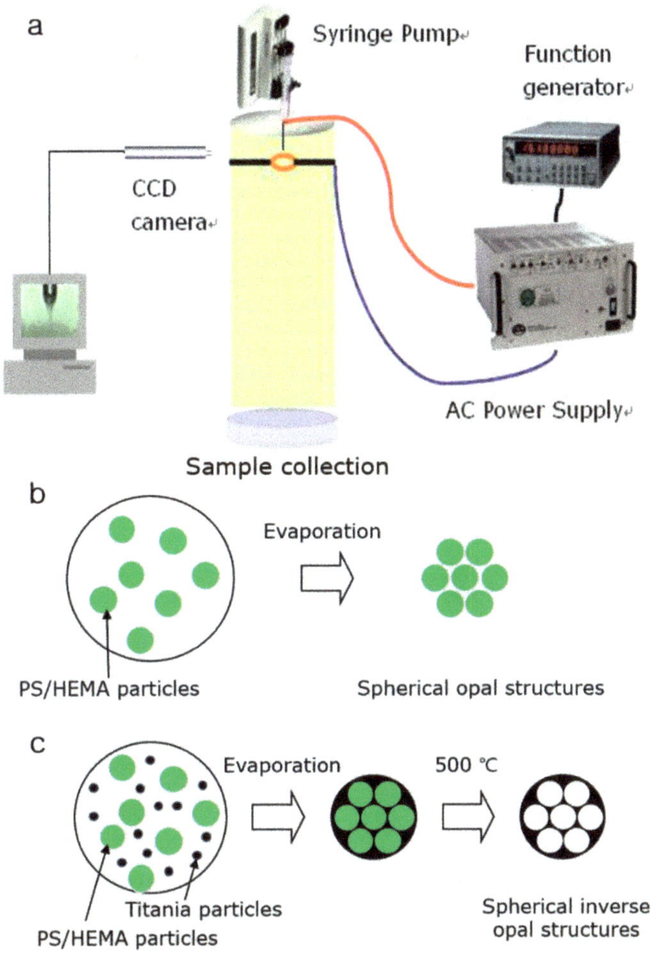

Scheme 1. (**a**) Schematic of experimental set up of electro-hydrodynamic atomization for generating uniform droplets. Schematics of self-assembly processes of (**b**) spherical PS/HEMA opal structures and (**c**) spherical titania inverse opal structures.

5. Conclusions

Uniform aerosol droplets could be produced by the Taylor cone jet mode of the electro-hydrodynamic atomization method using an AC electric field. As the solvent evaporated, PS/HEMA nanoparticles self-assembled in droplets to create a spherical opal structure consisting of PS/HEMA nanoparticles. When PS/HEMA nanoparticles and anatase titania nanoparticles were simultaneously used, spherical opal composite structure could be prepared. After removing PS/HEMA nanoparticles by 500 °C calcination, spherical titania inverse opal structure could be fabricated. Bragg's law was used to estimate the bandgap position (λ) of the spherical PS/HEMA opal structure and the spherical titania inverse opal structure. Since spherical opal structures could be prepared rapidly in large quantities using the electro-hydrodynamic atomization method, the bandgap position of spherical PS/HEMA opal structures could be visually confirmed from digital camera images. The bandgap position could be controlled simply by changing the size of PS/HEMA nanoparticles. The spherical opal structures and the spherical titania inverse opal structures produced by the electro-hydrodynamic atomization method are of practical significance for various applications, including reflective mode display, photo catalysis, solar cell electrode materials, and analytical systems.

Author Contributions: Conceptualization, J.-M.L.; methodology, J.-M.L.; software, S.J.; formal analysis, J.-M.L. and S.J.; investigation, J.-M.L. and S.J.; data curation, J.-M.L.; writing—original draft preparation, J.-M.L.; writing—review and editing, J.-M.L. and S.J.; visualization, J.-M.L.; supervision, J.-M.L.; project administration, J.-M.L.; funding acquisition, J.-M.L.

Funding: This work was supported by the Korea Institute of Energy Technology Evaluation and Planning (KETEP) and the Ministry of Trade, Industry & Energy (MOTIE) of the Republic of Korea (No. 20184030202130). This work was also supported by the Soonchunhyang University Research Fund.

Conflicts of Interest: The authors declare no conflict of interest. The funders had no role in the design of the study; in the collection, analyses, or interpretation of data; in the writing of the manuscript, or in the decision to publish the results.

References

1. Yablonovitch, E. Inhibited Spontaneous Emission in Solid-State Physics and Electronics. *Phys. Rev. Lett.* **1987**, *58*, 2059–2062. [CrossRef] [PubMed]
2. Holtz, J.H.; Asher, S.A. Polymerized colloidal crystal hydrogel films as intelligent chemical sensing materials. *Nature* **1997**, *389*, 829–832. [CrossRef] [PubMed]
3. López, C. Materials Aspects of Photonic Crystals. *Adv. Mater.* **2003**, *15*, 1679–1704. [CrossRef]
4. Arsenault, A.; Fleischhaker, F.; Von Freymann, G.; Kitaev, V.; Míguez, H.; Mihi, A.; Tétreault, N.; Vekris, E.; Manners, I.; Aitchison, S.; et al. Perfecting Imperfection—Designer Defects in Colloidal Photonic Crystals. *Adv. Mater.* **2006**, *18*, 2779–2785. [CrossRef]
5. Moon, J.H.; Yang, S. Chemical Aspects of Three-Dimensional Photonic Crystals. *Chem. Rev.* **2010**, *110*, 547–574. [CrossRef] [PubMed]
6. Kim, S.-H.; Lee, S.Y.; Yang, S.-M.; Yi, G.-R. Self-assembled colloidal structures for photonics. *NPG Asia Mater.* **2011**, *3*, 25–33.
7. Wang, J.; Zhu, J. Recent advances in spherical photonic crystals: Generation and applications in optics. *Eur. Polym. J.* **2013**, *49*, 3420–3433. [CrossRef]
8. Stein, A.; Li, F.; Denny, N.R. Morphological Control in Colloidal Crystal Templating of Inverse Opals, Hierarchical Structures, and Shaped Particles†. *Chem. Mater.* **2008**, *20*, 649–666. [CrossRef]
9. Kim, S.-H.; Lee, S.Y.; Yi, G.-R.; Pine, D.J.; Yang, S.-M. Microwave-Assisted Self-Organization of Colloidal Particles in Confining Aqueous Droplets. *J. Am. Chem. Soc.* **2006**, *128*, 10897–10904. [CrossRef] [PubMed]
10. Moon, J.H.; Yi, G.-R.; Yang, S.-M.; Park, S.B.; Yi, G.; Yang, S.; Pine, D.J. Electrospray-Assisted Fabrication of Uniform Photonic Balls†. *Adv. Mater.* **2004**, *16*, 605–609. [CrossRef]
11. Hong, S.-H.; Moon, J.H.; Lim, J.-M.; Kim, S.-H.; Yang, S.-M. Fabrication of Spherical Colloidal Crystals Using Electrospray. *Langmuir* **2005**, *21*, 10416–10421. [CrossRef] [PubMed]
12. Li, H.; Wang, H.; Chen, A.; Meng, B.; Li, X. Ordered macroporous titania photonic balls by micrometer-scale spherical assembly templating. *J. Mater. Chem.* **2005**, *15*, 2551–2556. [CrossRef]
13. Iskandar, F.; Nandiyanto, A.B.D.; Yun, K.M.; Hogan, C.J.; Okuyama, K.; Biswas, P.; Hogan, C. Enhanced Photocatalytic Performance of Brookite TiO2 Macroporous Particles Prepared by Spray Drying with Colloidal Templating. *Adv. Mater.* **2007**, *19*, 1408–1412. [CrossRef]
14. Hwang, D.; Lee, H.; Jang, S.-Y.; Jo, S.M.; Kim, D.; Seo, Y.; Kim, D.Y. Electrospray Preparation of Hierarchically-structured Mesoporous TiO2 Spheres for Use in Highly Efficient Dye-Sensitized Solar Cells. *ACS Appl. Mater. Interfaces* **2011**, *3*, 2719–2725. [CrossRef] [PubMed]
15. Zhu, F.; Wu, D.; Li, Q.; Dong, H.; Li, J.; Jiang, K.; Xu, D. Hierarchical TiO2 microspheres: Synthesis, structural control and their applications in dye-sensitized solar cells. *RSC Adv.* **2012**, *2*, 11629–11637. [CrossRef]
16. Chen, D.; Caruso, R.A. Recent progress in the synthesis of spherical titania nanostructures and their applications. *Adv. Funct. Mater.* **2013**, *23*, 1356–1374. [CrossRef]
17. Fattakhova-Rohlfing, D.; Zaleska, A.; Bein, T. Three-Dimensional Titanium Dioxide Nanomaterials. *Chem. Rev.* **2014**, *114*, 9487–9558. [CrossRef] [PubMed]
18. Taylor G, I. Disintegration of water drops in an electric field. Proceedings of the Royal Society of London. Series A. *Math. Phys. Sci.* **1964**, *280*, 383–397.

19. Lee, S.Y.; Kim, S.-H.; Hwang, H.; Sim, J.Y.; Yang, S.-M. Controlled Pixelation of Inverse Opaline Structures Towards Reflection-Mode Displays. *Adv. Mater.* **2014**, *26*, 2391–2397. [CrossRef] [PubMed]
20. Cardoso, A.H.; Leite, C.A.P.; Galembeck, F. Elemental Distribution within Single Latex Particles: Determination by Electron Spectroscopy Imaging. *Langmuir* **1998**, *14*, 3187–3194. [CrossRef]

Sample Availability: Samples of the compounds are not available from the authors.

MDPI
St. Alban-Anlage 66
4052 Basel
Switzerland
Tel. +41 61 683 77 34
Fax +41 61 302 89 18
www.mdpi.com

Molecules Editorial Office
E-mail: molecules@mdpi.com
www.mdpi.com/journal/molecules

www.ingramcontent.com/pod-product-compliance
Lightning Source LLC
LaVergne TN
LVHW070558100526
838202LV00012B/497